William Hewson

Principles and Practice of Embanking Lands from Riverfloods

William Hewson

Principles and Practice of Embanking Lands from Riverfloods

ISBN/EAN: 9783744712583

Printed in Europe, USA, Canada, Australia, Japan

Cover: Foto ©berggeist007 / pixelio.de

More available books at **www.hansebooks.com**

PRINCIPLES AND PRACTICE

OF

EMBANKING LANDS

FROM RIVER-FLOODS,

AS APPLIED TO

"LEVEES" OF THE MISSISSIPPI.

BY

WILLIAM HEWSON, CIVIL ENGINEER;

ASSISTED IN THE ENGINEERING REMARKS BY CONSULTATION WITH M. BUTT
HEWSON, ESQ., CIVIL ENGINEER, LATE CHIEF ENGINEER OF THE CENTRAL
RAILROAD OF MISSISSIPPI, &C., &C., &C.

Entered, according to Act of Congress, in the year 1858, by
WILLIAM HEWSON,
In the Clerk's Office of the District Court of the United States, for the Southern District of New York.

TO

THE HONORABLE JAMES L. ALCORN,

CHAIRMAN

OF THE SUPERIOR BOARD OF

LEVEE COMMISSIONERS FOR THE STATE OF MISSISSIPPI,

&c., &c., &c.,

THIS WORK IS DEDICATED

BY HIS FRIEND,

WILLIAM HEWSON.

NEW YORK, SEP. 15th, 1858.

To the Hon. James L. Alcorn,
 Of Mound Place, Coahoma County,
  MISSISSIPPI.

MY DEAR SIR :—

As father of the Levee-system, in at least that State, you are, *on public grounds,* the man of all others to whom ought to be dedicated the following results of my reflections and labors on the Levees of Mississippi. As a Legislator, as a County Commissioner, as a general Commissioner, for the conduct of those improvements, your zeal, energy, and talent, have contributed, in an eminent degree, to the present matured prospects of the most important material interests of your State—the drainage and reclamation of the great Valley of the Yazoo. As an agent, under your direction, in the accomplishment of that great result, I am bound, therefore, to lay at your feet this summary of the views and rules by which I have been guided in that agency.

As a *private friend,* however, you of all my friends in this country, have the first claim on my feelings. Truthful, warm, and disinterested, as I have ever found you in our long and somewhat trying intercourse, it affords me cordial pleasure to give public evidence, by even the dedication of this volume to you, of my profound conviction of your truth, warmth, and disinterestedness as a private friend.

While your qualities of head—in the capacity, the courage,

the integrity, with which you have managed the public interests committed to your care—command my unqualified respect, it is privately a pleasure to me—as publicly it is a duty—to inscribe to you, as I here do, this result of my experience and reflection on River-embankment. Accept it, my dear Colonel, as an humble expression of the honest regard of your ever faithful friend,

WILLIAM HEWSON.

P. S. Two years ago, this volume was commenced; though in the intervals of leisure occurring under active engagements of my mind, it had not been completed until to-day.

PREFACE.

IMMENSE public wealth is being accumulated behind the Levees of the Mississippi. From Cairo to the Balize, millions of acres of the finest land in the world are being reclaimed from the Bear and the Panther, to bring forth fruit for the enrichment of the American Union, and the luxury of private industry. Millions of money have already been expended on the works, behind which this great investment of enterprise—of labor and capital—is going on; and yet, up to the present hour, these expenditures have been made, to a great extent, without regard to the teachings of a regularly digested experience. This book—devoted to the Levee, and to Levee interest—is, therefore, given to the public, the *first attempt* to reduce to order and to rule, the design, execution, and measurement of the Levees of the Mississippi. Hundreds of thousands of people are concerned to the extent of their fortunes—if not, even of their lives—in the subject treated of here; and hence is it unnecessary for the author to apologize to the public, for an attempt to reduce a subject of such immense importance to an exact *popular rationale*.

The necessities of a country so new as that along the Mississippi, place the management of its public works in the hands of unskilled men. This fact, coupled with others proper to the case, commits the execution of the Mississippi Levees to the inexperienced good sense of the planter, or, less safe guide, to the inexperienced manipulations of the laborer, the wood-chopper, the mechanic, who succeeds in obtaining from that planter employment as a Civil Engineer. This condition of things, is, under all the circumstances of the case, to some extent unavoidable; and in order, therefore, to make it comport as far as possible with the proper management of the Levees, this summary of the principles and practice of Leveeing, is made in terms equally intelligible to the professional short-comings of those "Engineers," and to the popular knowledge of those planters.

Those gentlemen who, engaged on the Mississippi Levees are, *in fact*, Civil Engineers, will doubtless receive this humble volume favorably. They will, it is hoped, find something in it that may assist their labors directly, and much that will tend to strengthen their influence over the works, by convincing the intelligent planter, that those works are subjects of extensive reflection and experiment in a certain department of knowledge.

This Memoir, humble as it is in its pretensions, will do good service to the profession, and to the public, if it assist the parties interested in Levees to discriminate between the educated Engineer and the untrained pretender.

This volume contains a short review of the History of Embankments. Natural Phenomena of Rivers have also been considered in it briefly, especially those phenomena bearing more directly on the theory of Leveeing. The Engineer and Levee Commissioner will find these heads of some value for their own guidance, and also, in obtaining that co-operation for their plans, which must always follow from enlightening, on the subject of those plans, the population living behind the Levees.

The principles of Economy and Usefulness, touched on in this part of the subject, will be of great importance to the Engineer and to the Commissioner, in indicating useful reforms in the present system of Levee management and construction.

The Earth-work tables included in the following pages, will be found of general interest. They embody a new system of obtaining, by inspection, the areas of Earthworks in exact accordance with the prismoidal formula. They are alike applicable, as far as they extend, to the prisms of the Levee-bank, the Canal-bank or cut, the bank or cut of the Railroad. For slopes of a varying rate, or of greater extent than those given in the tables. the plan pursued in the preparation of those tables is equally applicable; and is, therefore, highly valuable to the practical Engineer, as a means of simplifying to an extraordinary extent, the laborious reference necessary in such tables as Sir John McNeill's, for calculating quantities by the prismoidal formula.

The Contractor on Levees is highly interested in the publication of the Earth-work tables included in these pages. Ignorance and carelessness have, too generally, characterized estimates of the quantities of Earth-work executed in Leveeing; the consequence being, sometimes, large losses to the contractor, sometimes large losses to the public. These tables, placing the facts within the reach of both parties, guarantee, therefore, justice in all cases to each. The use of the tables as explained in the letter-press, is short and simple. Men of ordinary intelligence, knowing the heights of a Levee, at intervals of 100 feet, can tell with accuracy by those tables the solid content of the Levee. The Commissioner, the Engineer, the Contractor, and the Public, can, therefore, bring these results within the compass of the popular knowledge.

CONTENTS.

CHAPTER I.
Sketch of the History of Embanking, ... 13

CHAPTER II.
Natural Phenomena of Rivers, ... 22

CHAPTER III.
The Levee, ... 53

CHAPTER IV.
Details of Levee Works, ... 79

CHAPTER V.
High Water Mark, ... 102

CHAPTER VI.
Location, ... 106

CHAPTER VII.
Surveys, ... 116

CHAPTER VIII.
Administration, ... 121

CHAPTER IX.
Earth-work Calculations, ... 134

The Tables 153

PRINCIPLES AND PRACTICE

OF

EMBANKING LANDS FROM RIVER-FLOODS.

CHAPTER I.

SKETCH OF THE HISTORY OF EMBANKING.

"LEVEEING"—Embanking as it is generally called—to confine rivers within their banks, and bar the approach of the sea, and its sister system of back-drainage, have, from a very early period, occupied the attention of individuals, governments and peoples.

The Phœnecians, Babylonians, Egyptians, Romans, Hindoostanees, and other East Indian nations, embanked low lands and drained marshes. Those nations chiefly inhabited alluvial plains, which, by their superior richness of soil when reclaimed, amply repaid them in the abundance of their crops, at less labor than was necessary to expend in the cultivation of higher districts. History informs us that the Babylonians and Egyptians were the first to adopt the system of reclaiming waste lands by embankments.

The low ground in the midst of which the city of Babylon was built, affords an early instance of the necessity of embanking; and consequently taught its inhabitants the principles of construction in earth works. The causeway thrown up over the low grounds, on each side of the Euphrates, leading to the celebrated bridge over that river, is the most remarkable, because it is the most ancient, of which there is any record.

Egypt, the land of floods and marshes, from the richness of its soil when reclaimed, was enabled at known periods of history to supply during times of dearth the impoverished nations around with corn. Egypt, when subject to Rome, was the granary from whence supplies for that city were drawn. About 2320 years before the Christian era, the greater part of Egypt was an extensive marsh, which Menés, the then-reigning King, undertook to reclaim. He diverted the course of the Nile into the middle of its valley or "bottom lands;" cut water-courses and raised embankments to confine the waters within them. His successors, each in his turn, made similar improvements—raised mounds on which to build their cities, above overflow, and cut canals for irrigation. The celebrated Lake Mœris is represented as one of the most remarkable works of ancient Eygpt; and is supposed to have been executed by a King of that name; and finished about 1385 years B. C. This Lake was according to Herodotus 450 miles in circumference, and is said by some to have been in places 300 feet deep. * By means of

---

* This summary of the great drainage works of the past follows without question the statements of History. Here, however, it may be observed that this Lake Mœris story seems to be one of the most preposterous inventions of even Herodotus. The pumping necessary to keep such an excavation dry is with even all the appliances of steam an inconceivable feat. The lifting and moving performances of the ancient Egyptians are truly wonderful, but how they could have hauled the infinite volume of excavation from such a "pit" as Lake Mœris, with an *average* haul of some 30 miles is beyond all comprehension. Waiving, however, these difficulties, the story is absurd. At an average depth of 150 feet—one-half the alleged maximum depth—the earth moved from Mœris would represent a content of 2,400,000 *millions* of cubic yards. Supposing the Egyptians furnished with the improved tools of modern excavations, this work at the rate of 5000 cubic yards per man per year, would have consumed for "loosening and lifting" the labor for 12 months of 480 millions of men. Done as stated in the reign of King Mœris, if it be supposed that he reigned for even

a canal it was supplied with water for six months in the year—this water during the remaining six months returning to the river by a regular system of irrigation throughout the whole extent of Egypt; thus supplying the land with moisture during the dry season. This Mœris *Canal*, in itself a stupendous effort of art, is still entire. It is 40 leagues in length. There were two others also in connection with the lake having sluices which were shut and opened alternately as the waters of the Nile varied in elevation.

The ancient Romans were more remarkable for the extent of their embankments, and the energy and skill put forth for the reclamation of submerged lands, than any other nation before their time. They appear to have been the special guardians of the swamps and marshes of Europe. The inducements to this guardianship lay in the great superiority and richness of those soils as compared with the more elevated districts.

The Romans embanked the Tiber near Rome, and confined the waters of the Po in a similar manner for many miles from its embouchure. Remains of their embankments are to be found in Holland, in most of the Fen-districts of England, and in other countries where their indomitable energy and perseverance carried them.

India presented originally great difficulties to culture. Its physical and its atmospheric character combine to present formidable draw-backs to natural production. Consisting largely of alluvial flats, the suddenness of its monsoon-rains, the short duration of those rains, and the long duration of the succeeding drought, place cultivation completely subject to these two conditions—irrigation at one season of the year, and drainage at another. Embankment is the prime means by which these

48 years, the work must have employed for that period in digging alone 10,000,000 of able bodied laborers! 50 millions of people must therefore have lived on the works, and a like number have been employed feeding them.

have been accomplished; and these embankments, or, as they are called in Hindostan, "Bunds," are the great artificial agents that have conferred the teeming luxuriance of the present state of that country on the soil of India. The "Bunds" of India, while damming back the floods of the monsoons from the rich flats behind them, run in double lines; and by confining between them the waters of those floods, reticulate Hindostan with numberless canals. Naturally, these waters would escape at periods of overflow, through a few rivers; but, distributed through numerous canals, they are retained on their passage to the sea sufficiently long to answer through the dry season the purposes of irrigation. To husband the too-copious monsoon-rains, the natives have built "Bunds" of great magnitude across river-vallies and streams, thus forming artificial lakes or reservoirs, often of vast extent, as storehouses, to supply the wants of a dry season. From these the water is conducted for miles along the flanks of mountains, across gorges and vallies, and through the most difficult countries, irrigating the land in its descent. Taught by the necessities of their country, the East Indian nations of by-gone ages have left behind them the remains of works of irrigation—monuments of their greatness—unsurpassed even by Egypt. One of these on the Island of Ceylon, an evidence of the enterprise and public spirit of the Cinghalese monarchs, is a good specimen of such works. It was formed of huge blocks of stone, strongly cemented together and covered over with turf, a solid barrier fifteen miles in length, one hundred feet wide at base, sloping to a top width of 40 feet, and extending across the lower end of a spacious valley.

Egypt was undoubtedly the cradle of the sciences, and particularly Hydraulics. It remained, however, for later times to arrange the laws of fluids into well defined formula. This science seemed to lay dormant for many hundred years: and it was not until during the eleventh and twelfth centuries, when it was thought necessary to make several of the Italian rivers navigable, and to cut canals for drainage and irrigation, and in the thirteenth century, when the practice of embanking and confining the rivers of Italy within their banks, was adopted, that it can be said to have claimed the attention of the learned. Before the seventeenth century there was scarcely any known digest of principles, by which, to carry out the works of the Hydraulic Engineer.

In the year 1665, a Congress of the most celebrated scientific men met in Tuscany. At this Congress it was proposed by Cassini and Viviana, to confine the Chiana by banks, and so conduct it to the Arno. During a subsequent meeting, at which Torricelli was present, the embankment of the Chiana was recommended on the ground that the rivers Arno, Tiber, and Po, were confined by the same means. At this period, practical Hydrodynamics received a great impetus; congressional meetings of scientific men were held, which, under the necessity of reclaiming all the submerged lands of Italy, called out the energies and talents of a host of the ablest philosophers of the age. The experience, experiments and writings, brought forth under these circumstances laid the foundation of our knowledge of Hydraulics in nearly all its branches. Amongst the distinguished men contributing at that time to this subject, were Gallileo, Torricelli, Guglielmini, Poleni, Manfredi, Zendrini.

A general system of embanking rivers, as a consequence of this movement amongst the *savans* of the day, was adopted in Italy, so that the Po, Adige, Tiber, Arno, Reno, and their tributaries are now confined between high, artificial banks.

Italy, from the peculiarity of its physical character, seems adapted by nature for the cultivation of this science. Lofty mountains, frequent rain-falls, heavy snow-drifts, break the region of the Alps and Appenines into, frequently, foaming torrents, torrents that, descending with headlong impetuosity to the more level country on the sea-board, pour destruction upon town and field. Along the low lands of the Po, perhaps the most fertile section of all northern Italy, from the destructive character of its precipitous floods, immense changes have taken place from time to time. The river has frequently changed its course, filling lakes and marshes, destroying towns, and causing immense devastations. Hydraulic Engineering has executed its first systematic works on the drainage of the Po ; and the nations of modern Europe have received therefrom, knowledge which enabled them to carry on similar works. France, Spain, Holland, Germany, England and Ireland, are thus alike indebted in this department of practical science to Northern Italy. And, now, embankments which reclaim immense bodies of rich land abound throughout all Europe.

Holland is well known to be low and flat. The alluvial deposits brought down, before even the dawn of History, from the higher districts of Western Germany and Northern France, by the Scheldt, the Meuse, and the Rhine, resulted in a marine swamp, known now as the "Low Countries." This once saltmarsh has been erected into the rich and prosperous Kingdom of Holland by "Dikes" or embankments. De Luck, in the first volume of his Geological Travels, says, " that the sea banks on the coast of the North sea, at the mouths of the Eyder and Elbe, extend to not less than 350 miles." And by another author, " the Southern shore of the North sea is embanked to the extent of 600 miles, the Southern shore of the Baltic for 1000 miles, and the Bay of Biscay to the extent of 300 miles." All the great rivers of Germany and Holland, such as the Rhine, the Elbe, the Oder, the Leck, the Vaert, the Yssel, the Maes, have all been confined to their channels by embankments.

The Zealand Dikes or embankments are said to be at least 300 miles in extent, and to cost for *annual repairs*, the large sum of $800,000 ! The sum expended for similar objects and for the regulation of the water-levels throughout Holland alone, amounts to $3,000,000 per annum ! The early history of embanking in Holland, Zealand, and other places, presents a series of calamities from the destructive power of water, almost unparalleled in history.

England is probably indebted to the Romans for the first embankments on the Thames. "London" indicates by its derivation from the Saxon words "Lyn din," that it was once "the City of the Lake." And history tells us that it owes its existence to the drainage of its site by embankments. The fact, however, may be settled without an appeal to history. It is well known that many of the marshes in the immediate vicinity of London—now under consideration, as subjects for embankment, are 12 feet below high tide in the Thames. The original marshy character of the ground on which the modern Babylon—like the ancient Babylon—stands, is indicated also by the fact, that many of its streets terminate with the word "Wall;" the names of several towns and places, such as Blackwall, Mill-wall, &c., on the Thames, are compounds of the same word, which in Kent, and Essex, is, to this day, the popular name for embankment. It is stated in an article published in the "Builder" of 22d August, 1857, headed, "Two Aspects of London;" that "All the space which is now so thickly covered with vast works, and occupied with living multitudes, was a watery waste, as desolate as the neighborhood of Babylon at the present day. Standing on a high part of Clerkenwell, or Islington, it is easy to imagine the picture;—a foreground of sedges, reeds, and willows. On the South East and West, a space of water extends to the base of the high-lands, presenting the appearance of a large lake in which the channel of the Thames is not even defined!"

The commencement of modern embankments in England, took place under Cromwell, about the middle of the 17th century. In 1478, however, the works undertaken by Bishop Morton, and subsequently completed by Charles the First, conjointly with the Earl of Bedford, and his friends, reclaimed 1,033,360 acres of rich land. In the space of a few years, previously to the year 1651, about 425,000 acres of fens, morasses, or overflowed lands, were recovered in Lincolnshire, Cambridgeshire, Hampshire, and Kent. Through the exertions of Sir Cornelius Vermuyden—a Zealander, who confined the Welland, and the Ouse within artificial embankments, a district has been reclaimed from the sea, in England, larger than the whole Kingdom of Holland. Sir John Rennie, in conjunction with Mr. Telford, constructed the celebrated Nenc-outfall, which, with the aid of banks, drained immense bodies of rich land. Mr. Wiggins says, that "the embankments on the coast of Essex alone, measure 220 miles." The principal rivers in England, subject to heavy freshets, are all embanked, the Thames, the Mersey, &c.

Of late years, embanking overflowed lands has been carried on extensively in Ireland; and in connection with the drainage of those lands by deepening the beds of the principal rivers and their tributaries, has been done almost exclusively by the Government, at the expense of the owners of the improved lands.

This summary of the historical facts of embanking is required, by the conciseness necessary in presenting the subject here, to be thus brief and general. Dugdale's history of embanking—an English work of great merit, will furnish the curious with the details of this great head of National industry; but sufficient has been here said, to indicate the extent to which the subject of Leveeing may be considered within the pale of practical science, and extensive practical experience.

Individual observation however extended will not, therefore, be brought by *really* intelligent men into conflict with the teachings of a knowledge contributed to by so many distinguished men, and tested by so many centuries of actual practice.

CHAPTER II.

THE NATURAL PHENOMENA OF RIVERS.

Levees on the Mississippi are both important and costly. Works involving so much public and private interest, and so much public and private outlay, ought to be predicated on principle. To oppose the laws governing a mighty river, is a labor from which even the Hercules of American energy may well recoil, and therefore does it become a duty of common sense to place that energy, in dealing with the Mississippi, at a labor in the least possible discord with those laws. To do this, it becomes necessary in the first place, to study the habits of rivers generally, and of the Mississippi particularly when its habits are separated from those other rivers by specialities. The particular habits of the Mississippi may be made subjects of local observations; but in order to confine the sphere of those observations to its proper limits, and to assist in its inferences, it is necessary to consider here in popular terms, the general rules affecting the régime of rivers as applied specially to the Mississippi.

Science is acquainted but generally with the causes by which river phenomena are influenced or the complicated laws by which they are governed. The little success that has attended the labors and reflections of enquirers on this subject from the time of Gallileo, is attributable to the difficulty of making correct observations, and to the local specialities which exist in most rivers. The following review contains a summary of the exact laws, and approximate rules deduced from observations of rivers.

The surface of a country may be generalized into a series of inclined planes—those planes ascending from the sea-level at the shore, to the mountain-heights of the interior. In a paper drawn up for the Institute of Civil Engineers by Mr. M. Butt Hewson, explanatory of the system pursued by him in carrying out certain works for the Board of Public Works in Ireland, the hydrographical distributions of rain-shed are thus indicated: "The surface of a country is resolved by its drainage-waters into several systems of vallies. These vallies are termed by Engineer's rain or 'catchment' basins. The line of lowest level in each of these is traced by a stream; this stream tends to its debouching point, more or less tortuously, more or less inclined. A catchment-basin is generally resolvable into several minor vallies associated together by the direct discharge of their respective streams into one common outlet; the several points of this discharge marking the several stages of increase in the area of the basin. A great central valley traverses the lowest level of these river-basins, secondary vallies branching from each side of this, constituting in their turn central vallies to distinct portions of the whole catchment. Tertiary vallies—so to speak—branch again from these secondary vallies, and like those secondaries, become so many distinct trunks to so many distinct systems of branch vallies. And so on until the sources of mighty rivers are traced on all sides of their basins, into the ravines of far off ridges, and the gorges of snow-capped mountains." The physical features of a country being of this character, the fact of rain-falls results necessarily in the facts of Cascade, Rapid, Stream, Rivulet, and River. Drawn up by that great mechanical agent, Heat—evaporating water from the surface of the earth and of the sea—water-laden clouds, driven and distributed by the winds, are by various causes precipitated on the land; and bursting over any portion of a country their waters are congregated by gravity from the higher into the lower vallies, and collecting strength as they rush down their several inclined planes, push forward in their downward course, over cataracts, down rapids, through lakes, and gently sloping streams, till uniting in one common grand volume in the primary valley, they roll forward to the sea, in the power and magnificence of a Mississippi, an Amazon, a Ganges, or a Nile. The amount of water thus falling upon the earth is not less than thirteen hundred millions of gallons per second throughout the year; one-half of this quantity is believed to run off the surface of the earth directly in rivers, which is the cause of floods, one-fourth is evaporated and taken up by vegetation, and the remaining fourth passes into the earth, keeping up a constant supply to the numerous springs which, to a great extent, feed and preserve the summer flow of water-courses.

Large and small rivers are governed by the same laws, under the same circumstances. The smallest rivulet has its own catchment-basin, or rain-shed, corrodes the bank that confines it, and pushes forward towards the sea, in proportion to its strength, the matter thus detached and held in mechanical suspension by the rushing of its waters. This smallest rivulet is characterised by its overflows, its sand-bars, eddies, sinuosities, and siltings-up of bed along its course, and at its mouth, in precisely the same manner as the great "father of waters," or any of the other mighty water-courses of the earth.

The course of all rivers is so devious that the distance between their extremities is very frequently twice the length of their rectilinear distance. Every obstacle or projection in the bank where the soil is harder and of a more resisting nature, the slightest irregularities in the bottom and sides, partially obstruct its course; and according to the magnitude of that projection, deflects, or tends to deflect, the current to the other side. This deflection of the current produces a circular motion in the water, which acting on the soft portions of the bank, hollows it out, forms eddies, and accelerates change in the direction of the current.

Overflows or freshets, in rivers are very variable in volume. They are dependent upon a variety of causes being brought to bear to produce them. Heat and cold, clouds and winds, forests and mountains, as first causes, are all intimately connected with their origin; cultivated lands, dense forests, water-bearing strata, and rocks of a permeable and of an impermeable nature, as secondary causes, have each their respective influence in passing off rapidly, or in passing off slowly through springs, the waters falling upon them. The depth of rain-fall varies greatly in different hydrographical districts, so that two rivers with the same extent of rain-basin, may differ largely in the amount of their maximum volume. Rain-shed is greater in mountainous districts than in plains. It is greater in equatorial than in polar regions, and varies, even in the same latitudes, to an extent often as great as that between Northern and Southern districts; for instance, it is greater in Ireland than in Russia, and it is greater on the Western slopes of the Cascade and Rocky mountains, than on their Eastern slopes. The sudden melting of snow, or a continued rain-storm, will sometimes congregate rapidly into a river channel an amount of water equal to a high multiple of its average outflow. At Marseilles, France, in a shower of rain of 14 hours duration, thirteen inches have fallen, constituting a high proportion of its rain-fall for twelve months; and at London, England, six inches have been known to fall in one and a half hours—nearly one-fourth its mean annual fall. And from the same cause, Western rivers are seen occasionally to rise from 15 to 20 feet in 24 hours; and even a height of five feet, like a wall, is sometimes observed to come rolling down, sweeping all before it in the descent. There are various causes in the river-bed, acting to retard the flow of those waters, and helping by that retardation to raise their surface, such as friction, eddies, sinuosities, and other circumstances. M. Venturi deduces from his experiments on tubes with enlarged parts, "That eddies destroy part of the moving force of the current of the River, of which the course is permanent, and the sections of the bed unequal, the water continues more elevated than it would have done, if the whole river had been equally contracted to the dimensions of its smallest section, a consequence extremely important in the theory of rivers, as the retardation experienced by the water is not only due to the friction over the beds, but to the eddies produced from the irregularities in the bed and the flexures and windings of its course, a part of the current is thus employed to restore an equilibrium of motion which the current itself continually deranges." The irregularities of river-beds, and the irregularities of rain-falls are thus seen to be combined in producing the phenomena of floods. The irregularities of rain-fall are of course causes beyond human influence, but the co-operating cause of floods—the peculiarities of river channels—are within the field of human operation, and become, therefore, the special object of enquiry to the Engineer engaged in, and people affected by permanent or periodic overflows.

All rivers decrease in their rate of descent as they approach the outfall; and this decrease is made over a series of curves gradually flattening until they flow out into the tangent or horizontal line of the sea-level. Short water-courses and minor river-basins in mountainous districts are generally of a precipitous character. Such are those of the Alps, and of the Western slopes of the Rocky mountains. The large rivers with which this continent abounds are generally for the greater part of their length of slow descent. The average fall of the Mississippi river for the whole distance from the gulf of Mexico to the confluence of the Ohio, following the windings of the river during *low water*, averages very nearly three inches per mile. Supposing the channel of the river straight and the rate of descent uniform between those points, the fall would be about six inches per mile. But the actual rate of declivity in a river is considerably increased in times of flood. Thus, the Nile falling into a tideless sea, rises, at the city of Cairo during floods, 25 feet, at Thebes, 36 feet, and at the first cataract—a point nearly equally distant from the mouth as Cairo from the Balize—about 40 feet. The Mississippi river falling into a sea too of a very nearly constant elevation—rises at New Orleans 12 feet, at Friar's Point, Mississippi, 42½ feet, and at Cairo, Illinois, about 50 feet. This flood-rise at Cairo, added to the elevation of *low* water at that point, gives an average rate of fall for *high* water, following the windings of the river, of three and a half inches, and on a direct line, of seven inches per mile. There is, therefore, in the circuitous and in the rectilinear distances a difference of fall equal to half an inch and to one inch respectively per mile, due to the average rate of fall during *high* water at Cairo, over and above that due to the average rate of fall from the same point during *low* water.

A fixed expression has been deduced for the velocity of small conduits. No formula for the purpose is yet found, nor is one likely ever to be found, applicable to the infinitely varying conditions of velocity in large rivers. Whatever may be the co-efficients and the combinations necessary to give preciseness to any mathematical expression for river-flow, the terms of that expression may be held, in general, to be the rate of fall, the depth of volume, and the content of the sectional area or friction-surface of the bed. Any change in either of those three conditions, will—all things else being equal—involve a change in the velocity of a river. Gravity being the motor in all cases of water-flow, that flow would, unless under the influence of some retarding cause, take place, like the free fall of any other body, with a constantly operating acceleration. These retarding causes are more than one in the case of river-flow; they are represented by the loss of mechanical effect arising from the shocks of the bank, and the friction of the bed. The retardation arising from friction is one of constant operation. The lower planes of the cross-section of river-flow suffer more active retardation from this cause than the upper planes; the former being retarded by the friction of their motion over the roughness of the *bed*, while the latter are retarded by the greatly reduced friction of sliding over the comparative smoothness of the *lower* water-planes. The deeper the volume of a river, the higher therefore, is, not only its surface velocity, but also, its mean velocity. And the same remarks apply to the *vertical* planes of the volume; the greatest retardation taking place at the sides, the least in the centre of the stream. The greater the width, therefore, the more active—so far as side-friction can effect the result—is the flow. But the friction of the sides is so small as compared with that of the width, that the latter general deduction may be disregarded; and we may conclude, practically, that an active increase in the velocity of a stream—a decided diminution in the retardation of friction—is always the result of an increase of depth.

In even small streams a fall of one-tenth of an inch per mile will produce a sensible flow. In large streams this rate of inclination, would, as seen by the above reasoning, result in a current proportionally more considerable. The frictional resistance of a river-bed is higher for higher velocities than for lower; varying, according to the observations and deductions of M. Eytelwein, as the rate of the square of the velocity. Such are the general facts of fall and friction.

The effects of tributary waters, on the volume and velocity of streams, appear somewhat paradoxical. Gennétte, supported by M. Eytelwein, asserts that one river may absorb another of equal magnitude with itself, without producing a sensible elevation of its surface. Cressy, in his Encyclopedia of Civil Engineering, sustains this opinion by citing the absorption of the Inn, by the Danube; of the Mayne, by the Rhine; of the Sechio, by the Po, and of the Teverone, by the Tiber; this absorption, he states, taking place without making the volume of the absorber in each case either deeper or wider. The only effect of the accession to the body of water passing through the main channel, in each of the instances named, is said by Cressy, to be an increase in the velocity. Guglielmini, in evidence of the same opinion, refers to the accession of the waters of the Ferrara, and Panaro, branches of the Po, to the volume of that river without, as he alleges, producing any sensible augmentation of its channel.

A corresponding increase of velocity must of course be supposed a consequence of such an accession, if we are to accept as a fact, that the accession of a tributary has no effect on the width or the depth of its main out-fall. This increase of velocity in the united volumes must, however, be referable to some commensurate mechanical cause. The tributary volume, it is true, discharges into the united volumes, the velocity proper to itself; and therefore, waiving the fact of altered rate of fall, or of altered depth of flow, the united volumes may be held, on mechanical grounds, to flow, after union, at a proportional *average* of their respective velocities. That the union of the two waters flows off without any increase in the original volume of the main stream, were to suppose the result of their blending of mechanical effect, the *sum*, volume for volume, of their original rates of flow. If the two volumes were, for example, equal, the one moving originally at *two* miles an hour, the other at *three* miles, then would the discharge of the united volumes, without increase of width, or depth, or rate of descent, suppose the resulting velocity to be *five* miles an hour. Mechanically this is not supposable. Therefore, must we come to the conclusion that it is impossible that the union of two rivers can take place without an increase after the union, in either width or depth. Eytelwein and Cressy must clearly have either mistaken the fact or have stated it erroneously.

The conclusions of the respectable names given under this

head, are, like all conclusions, open to question. The *facts*, however, must be received beyond all doubt. While there can be no question as to the facts, that the Danube, the Rhine, the Po, the Tiber, in all the instances of accession named, have not been widened or *elevated ;* the inference is irresistible, that in all these instances, they must have been, to at least some extent, deepened. The fact of deepening, resulting, as premised above, in a proportional diminution of frictional resistance to flow, involves directly an increase in the rate of flow. This, combined with the mechanical impulse of the tributary volume, must, by accelerating the velocity, make the increase of depth proportionally less than the increase of accession.

It has been remarked by several writers that the width of the Mississippi below the junction with the Ohio, is less than its width above the junction. This is not only true of the river in the case of the accession of the Ohio, but also, of all accessions below that, and indeed, of the channel generally from Cairo to the Balize. At Cairo, the Mississippi is upwards of a mile wide; at New Orleans, the width is but half a mile. But this narrowing down-stream is accompanied by a corresponding deepening—a truth that is established popularly by the fact that the higher a steam-boat goes up stream, in low water, the more difficult is the navigation; until, at Cairo, further navigation at such times becomes almost impossible, even for the smallest craft.

A rough approximation of the sectional areas, in times of flood, of the Mississippi, at Cairo, and at New Orleans, in conjunction with a like statement of all its intermediate tributary streams, will be found on the next page.

An accession of some 500,000 square feet of tributaries is seen by this statement, to be passed through the Mississippi river at New Orleans with an increase of volume over that at Cairo, of but 160,000 square feet; and through a channel upwards of twice the depth, and but one-half the width.

At Cairo, the sectional area of the Mississippi, is about,	325,000 feet.
Of the Ohio, at junction, the sectional area is about,	260,000 feet.
Of the St. Francis, at junction, do.	21,000 feet.
Of White River, at do. do.	28,000 feet.
Of Arkansas, at do. do.	56,000 feet.
Of Yazoo, at do. do.	21,000 feet.
Of Big Black, at do. do.	21,000 feet.
Of Red River, at do. do.	52,000 feet.
Of other tributaries, at do. do.	18,000 feet.
Total,	802,000 feet.
Of the Mississippi, at New Orleans,	480,000 feet.

This fact, ascertained loosely as it is, establishes the correctness of the general conclusion reasoned to above, namely, that while on the authority of the statements of Gennétte, Eytelwein, Guglielmini, Cressy, we must accept the fact that tributary accessions to the volume of a river do not widen, or elevate their general level, all such accessions result in an accelerated velocity, and an increased depth. As a practical application of this conclusion in the case of the Mississippi river, it may be, therefore, safely affirmed, that the retention of flood-water in the channel by levees, like all tributary accessions to its volume, while deepening the channel, and increasing the velocity will not, as a *direct* consequence, elevate the surface of the water.

The conclusion arrived at in the foregoing paragraph appears on its face paradoxical. Paradoxical or not, it must be observed that it is a conclusion drawn fairly, from undoubted premises. It will be said, if the enclosure of *surface*-flood-water within the channel do not elevate the level of the river-flow, how is it that the accession of *any* flood-water at all produces that elevation? The inference drawn above is not affected by this question; because, not declaring that there are *no* variations of river-level, it applies to only those circumstances under which a tributary-flood is discharged into the river-channel at *the period* of a corresponding flood in that main channel.

from its own supplies. The conclusion arrived at is in truth this :—a glut of water in the Mississippi will not be increased in level by the accession of other gluts, from the Ohio, Arkansas, &c. But if we are to suppose every accession of floodwater an accession of height; and that we begin with over-flows of even 6 feet at the accession of the Missouri, of 6 feet additional at the accession of the Ohio, of 6 feet more for all accessions of minor streams, of 6 feet more for the accession of the Arkansas, of 6 feet more for the accession of Red River, the flood level at New Orleans—assuming no adaptation of channel as we go down-stream—would be 30 feet above the surface of the land ! But what, on the contrary, is the fact ? The elevation of floods at New Orleans is altogether but 12 feet above the low-water mark, which *increasing up-stream*, it is in fact, at Cairo, 50 feet—and this in the face of all the accessions from Hatchees, St. Francis, White, Arkansas, Yazoo, and Red River.

The direct agent of change in a river-course is the current. On the banks this acts in two ways—by friction, and by impact. The greater the velocity the greater of course will be the length of the rubbing body that, moving along the bed and bank, constitute the friction. The friction, therefore, varies with the velocity; being twice as great for two miles as for four miles. Friction, varying also, as the weight of the rubbing body varies as the depth, being twice as great in a depth of 40 feet, as in a depth of 20 feet. All sections of channel are subject to this consequence of flow; but the more even and regular the section, the less the friction. In irregular and uneven sections the friction runs from friction proper into impact.

Impact begins in channels where friction ends. A stream flowing over a smooth, straight bed is resisted by only the adhesion due to friction; but over a rough, crooked bed is resisted, in addition to this adhesion, by shocks to the regularity of its flow, whether against shoals, bars, stumps, or bends. This further resistance combines within it all those impediments involving impact; and for whatever part of the cross-section of the flow is engaged in this impact, varies as the weight of that cross-section multiplied by the square of the velocity. The weight, however, varies directly as the depth, being twice as great for a depth of 50 feet, as for a depth of 25 feet; and hence does a river become the most powerful agent of change by impact, at periods of highest flood. The velocity, too, increasing with the depth, shews again and in a higher degree, why a river exerts its greatest energy, so far as impact expresses that energy, at the period of its greatest depth. For impact as measured by velocity increases as the square of the velocity, being nine times as great for the same impediment and the same depth, in a stream of six miles an hour, as in a stream of two miles an hour.

The effects on river beds and banks from friction and impact, cannot be given here more satisfactorily than in those general elementary terms. No experiments that have come under my knowledge, furnish a measure of the effects of friction and impact, in the case of rivers, by practical examples.

Friction and impact, so far, have been touched on as agents for *excavating* material. After this excavation, however, they continue to act on the material excavated with their combined forces. A lump of earth for example, being rubbed off by friction or knocked off by impact in the channel, is taken up by the water and impelled forward by the rubbing and the striking of the flow. Small bodies, and bodies of a weight a little more than water, are thus moved along by the stream in suspension; larger bodies of a weight considerably greater than water, being, by the same power, rolled forward over the bottom. This energy, this power of transportation of material within its channel by a river, may be understood in relative terms by the remark that it is the combined effort of friction and impact—of the rubbing of the planes of water that flow past that material and of the striking of that part of the flow which it impedes. Practical results, however, give this energy a plainer expression. The following facts, ascertained after a series of careful experiments by Dubuat, show clearly the absolute energy of several velocities of rivers for the transport of materials loosened by their currents, or otherwise deposited in their beds:—

Clay fit for pottery removed by water flowing at the rate per second of	3¼ inches.
Fine sand removed by water flowing at the rate per second of	6¼ inches.
Gravel about the size of peas removed by water running at the rate per second of	7¼ inches.
Gravel about the size of beans removed by water running at the rate per second of	12¼ inches.
Shingle—large gravel—about one inch in diameter, removed by water running at the rate per second of	25¼ inches.
Flints about the size of hen's eggs removed by water running at the rate per second of	40 inches.
Broken stones removed by water at the rate per second of	48 inches.
Soft rocks begin to yield with a velocity per second of	52 inches.
Rocks with distinct stratification begin to yield with velocity per second of	72 inches.
Hard compact rock begins to yield with a velocity per second of	120 inches.

From this table it appears that the very moderate velocity of 950 feet per hour, is capable of moving clay; of 1900 feet per hour, capable of moving fine sand, and of half a mile an hour, capable of moving coarse gravel. The carrying or propelling power of a stream on bodies within it, is seen from the table to increase with its velocity; the materials capable of movement in a current of 4 miles an hour, being incapable of motion at 3, 2, or 1 mile an hour. This fact leads to some of the most important changes in rivers as will be shown below. As no practical examples of the abrasive effects of a current have been given above, it may be observed here, that those effects, resulting as they do from the same causes, which, certainly with an energy less in degree, constitute the propelling power of currents, are presented in the facts of the above table *relatively*.

Impact and friction "washing away" or "caving in" the material of a river channel, it has been seen that impact and friction continue to act afterwards on the material so "washed" or "caved," for its propulsion along the channel into the outflow or sea. It has also been shown that this propelling power is greater or less as the velocity of the stream is greater or less. Friction, it has too been premised, is greatest in its retardation of flow at the bottom and at the sides, the rate of flow being always greatest in the middle, and at the top of the stream, and diminishing from that top and from that middle on either side, until, at the bottom and at the sides, it becomes the least. This consideration may be put in this shape: the velocity of a stream is unequal at every point of its section. Now a mass of material broken off or rubbed off by a current from the bed or bank is, when so broken or rubbed off, placed free to move in a current of a certain velocity. This velocity, let it be supposed, has energy sufficient to propel that mass. Like all moving bodies, the motion of this mass tends to follow a straight line in the direction of the force acting on it. But at a sudden bend or obstacle in the river the thread of the stream, acting on this soft mass, may be supposed to be deflected at a large angle with its original deflection. The mass, being heavier than the corresponding thread, will be deflected from its original direction at a *smaller* angle than the water; and hence, will take its place in a different part of the water-section. By irregularities in the direction of the flow the bodies rolled along a river bed are thus seen to be constantly shifted from one position in the cross section to another position in that section; and consequently are seen to be *shifted into velocities constantly changing*. Of bodies carried along by a stream and so light as to be held at first in suspension, the constant action of gravity tends to the *depression* of those bodies; and thus, in urging them constantly downwards, brings them in their dropping through different depths, into different rates of flow. All material then, whether large or small, light or heavy, are seen to be subject in river channels, to constantly changing velocities; and hence, the *carrying power* of rivers depending on velocity, all material too heavy or too large to be moved by the *smaller* velocities of the cross-section of the stream, when once shifted into a position having any of those smaller velocities, sinking to the bottom, becomes fixed. The lighter and smaller bodies rolling along that part of the section where this material thus becomes fixed, accumulate around it as a nucleus, and this process of fixing heavier matter and retaining lighter matter, results in shoals, bars, islands, and those deposits known on the Mississippi, as " *making banks.*"

The amount of material carried down-stream by a river, varies, as has already been seen, with the velocity and volume on one hand, and varies on the other hand, with the hardness or softness, lightness or heaviness of the material composing the bed and bank. The quantity of solid matter borne forward by the Ganges is estimated at 1-40 of its volume, the total quantity of earth propelled per year by that river being estimated at the almost incredible amount of about 315,000 million cubic yards. The Rhine is estimated by Mr. Horner to propel solid matter to the amount of 1-16000 of its volume. At New Orleans the earthy matter propelled by the Mississippi is estimated by Dr. Riddell, taking a mean annual average, at an amount of 1-1700 of the volume of the flow. The Mississippi is shown by the estimate of Sir C. Lyell to carry earth below New Orleans to the amount per annum of 137,000,000 cubic yards.

Of the whole material propelled by a stream, a proportion has been seen to be *precipitated* from mechanical causes in the form of shoals, bars, islands, "making banks," &c. The residue, however, of this quantity of matter is carried forward to the debouch; and pushed for further propulsion into the outfall stream, or partly for further removal, partly for permanent deposit—impelled into the sea. This deposit of material by sea-discharging rivers, is the cause of that general accompaniment of a system of rivers—the Delta. Constant in its operation, this cause of the formation of Deltas would, under circumstances always the same, lead to the constant extension of those Deltas. This extension, however, must be held under its general circumstances to take place, as measured by its direct advance, at a rate constantly diminishing, until finally it shall have reached its limit of direct advance. The trend of a shore-line may, for instance, be supposed to place the debouch of a river in dead water; and this dead water, favoring the precipitation of material, the earth propelled into it by the river, produces, to a certain point, a constant direct advance of that river's Delta. At this certain point, however, the Delta may be supposed to have passed from the dead water of its original formation, and to have become subject to the disturbing influence of an active current. Direct advance at this stage of its growth may be thus considered at an end. The direction of the river-flow crossing the course of the sea-current at an angle, the resulting direction of commingling and of deposit, follows a bend increasing more or less rapidly, according to the energy of the sea-current, as compared with that of the river, until, finally, it shall have assumed the line of the sea-current. Such, in general, are the causes and condition of the growth of Deltas. They apply alike to the Deltas of the Mississippi in this country; of the Orinoco in South America; of the Ganges, of the Irrawaddy, of the Indus, &c., in Asia; of the Nile, and of the Niger in Africa; and of the Rhine, of the Rhone, of the Po, of the Danube, &c., in Europe.

The rates of advance in Deltas, consequent as they are on the varying causes affecting their formation, are variable for different Deltas.

The Ganges and Burrumpooter, deliver into the Bay of Bengal solid material to the enormous amount—during the flood season—of 500 million cubic yards every twenty-four hours. There is no information at hand, as to the rate of growth of the Ganges-Burrumpooter Delta. The incompleteness of the recorded facts of the rivers of the great Deltas of the world, makes it impossible to deduce any general law as to the rate of Delta growth in any particular case. In this place may be added all that are at hand of the facts of Delta growth; and, indeed, perhaps this may, after all, be quite sufficient for the practical purpose aimed at under this particular head.— The Delta of the Nile has advanced but two miles since the time of Herodotus; but small as the consequent rate of advance is, it has now been ascertained to have altogether ceased. The Po, and the Adige, discharging at the same point into the Adriatic, have formed their joint-Delta since the time of the birth of Our Saviour. One hundred miles in width, this Delta has, up to the present time, advanced into the sea upwards of 20 miles. Sir C. Lyell, after comparing the present tongue of land below New Orleans, with the map published by Charlevoix, alleges that the Delta of the Mississippi River has not advanced more than a mile in a century. Mr. Rogers, in his report to the British Association on the Geology of North America, says however, that, "as an example of the rate at which it is growing, the old Balize erected at the mouth of the river, about the year 1724, is now (1834) two miles above it. There was not at that time, the smallest appearance of the island on which, 42 years after, Ulloa caused barracks to be erected for the pilots, and which is now known as the new Balize. The distance from the mouth of the river at which the chief deposit of sediment usually takes place is about two miles; when these shoals accumulate sufficiently they form small islands, which soon unite and reach the continent, and thus the Delta increases."

In this statement of the growth of Deltas it must be observed that the statement for that of the Nile and of the Mississippi, is applicable to rivers unconfined by Levees. The Nile overflows its banks without artificial restraint. The Mississippi, up to the period of the observations referred to, had been but very partially Leveed; and hence do those observations of Mr. Rogers refer to a river without Levees. The facts of the growth of the Delta of the Po-Adige are, however, since the sixteenth century, those of a Delta formed by a river whose floods are confined within artificial banks. The rate of advance of the Delta of the Nile from the birth of History until now, has been 4 feet a year; of the Mississippi from 1724 to 1834 has been 96 feet a year; of the Po-Adige, for the period between the beginning of the first and the beginning of the thirteenth century, 22 feet a year; for the next following 400 years the advance has been 82 feet a year; and for the 200 years next after that, it has advanced at the rate of 229½ feet. The present Levee system of the Po had its origin in the 13th century, but was incomplete until the commencement of the 17th century. Since the beginning of the 17th century, however, the embankments of the Po and Adige have been completed from end to end. The unleveed period of the Po shows an annual rate of advance in its Delta of 22 feet. But from the introduction of the Levee-system on that river (taking the average during the whole period of its progress,) the rate of advance of the Po-Adige Delta ran up from 22 feet annually to 82 feet: and from the completion of the Levee-system, taking the experience of 200 years, the advance of the Po-Adige Delta has run up from 82 feet annually to a yearly rate of 229½ feet. The conclusion then from the experience in the case of the Po is irresistible, in the absence of any other especial cause, to account for such an accelerated advance, that the confinement of the river Po within embankments has caused its Delta to advance into the sea with comparative rapidity. Levees therefore, may be held to involve an accelerated rate of extension of a river-Delta.

The advance of its Delta exerts decided influence on the high-water level of a river. The flood-height of the Mississippi, which at New-Orleans has been stated already at 12 feet above low-water, is at Friar's Point $42\frac{1}{2}$ feet above low water, and at Cairo is 50 feet. Every 3 inches of elevation at New Orleans represents therefore an elevation at Cairo of $12\frac{1}{2}$ inches. Now the rate of fall from New Orleans to the sea is about $1\frac{1}{2}$ inches per mile, and therefore an advance of the Mississippi Delta at an accelerated rate based on the acceleration resulting from Levees to the advance of the Po-Adige Delta would give—by an extension in 100 years of $4\frac{1}{3}$ miles of Delta—an additional elevation of $6\frac{1}{2}$ inches to flood level at New Orleans, an additional elevation to that level at Friar's Point, of 23 inches, and at Cairo an additional elevation of 27 inches. The relative height of high water at any point on the Lower Po, in comparison with that at any point on the Upper Po, is not conveniently obtainable; but assuming it the same as between that at New Orleans, and that at the Balize on the Mississippi, the extension of the Po-Adige Delta since the completion of the Po and Adige Levees— 9 miles of extension—must have occasioned, for the preservation of the same rate of incline of outflow from Ferrara down stream, as from New Orleans down stream, an elevation at Ferrara of 13 inches. The elevation of the Po, however, at Ferrara is measured not by inches but by feet; and the increase of this elevation since the completion of the Levees must, therefore, be referred to some other direct cause than the extraordinary extension of the Delta.

The overflow of a river discharges a large proportion of its earthy matter upon the land. The confinement of the River within Levees confines this proportion of its earthy matter to the channel. The immense amount of the material so added to the work of the stream, may be inferred generally from the fact that in the case of the Nile, it was distributed over Egypt by overflow, and has caused the elevation of *the whole surface of the country* since the Christian era, at an average rate per hundred years of 4½ *inches*. The greatness of the aggregate mass of matter added to the original proportion in its volume by the construction of Levees, may be inferred generally, by the immense additions resulting from Levees to the growth of Deltas. But the carrying power of a water-course, like all other mechanical agencies, has its limit; and when we see any cause loading it beyond its previously established energy, we may reasonably expect that a portion of its excessive work will of necessity be left undone. The motive power of a river acting up to its limit in the removal of matter from its source to the sea, may be readily supposed under its insufficiency for the removal of the extra matter accumulated within its Levees to drop a portion of that matter into irregularities in its bed. The matter so dropped may be supposed to accumulate in layers, as every accession of material increases the weight of matter to be moved, over and above the energy of the stream. But these causes of deposit in the beds of rivers apply in the surcharging of matter in streams whether Leveed or unleveed, though from the retention of *all* the matter within the channel by Levees, much more strikingly in the case of Levees. The Nile illustrates the fact that unleveed rivers undergo a constant elevation of their beds; for while the matter deposited during the overflows of that stream as already stated, has elevated the surface of Egypt 4½ inches per century, the matter deposited within the bed of the river has elevated the level of that bed at the same rate. The facts in this case are so well defined that it may be well to place them here on record. At Damietta, the Balize of the Nile, where the elevation of overflow in the river is imperceptible, the elevation in the level of the river-bed and river-bank is inappreciable. At Cairo, 120 miles from the mouth, where the flood-level is 25 feet above low water-mark, the elevation of the land and of the river-bed is, since the Christian era, 5 feet 10 inches. At Thebes, 500 miles from the mouth, where the flood-level is 36 feet above the low water line, the land and the river-bed have been elevated, since the birth of Our Saviour, 7 feet; while at the first Cataract, 100 miles higher up-stream, the level of the bed and of the bank have been raised, since the same period, as much as 9 feet. Assuming the same width of channel in the Nile at Cairo, at Thebes, and at the first Cataract, and assuming further the same amount of detritus carried off, volume for volume, by the overflow at each of those three places, we may not be surprised to find that a 40 foot flood, giving an elevation of bed to the extent of 9 feet, a 36 foot flood an elevation of bed to the extent of 7 feet, and a 25 foot flood an elevation of bed to the extent of 5 feet 10 inches, the height of flood bears an almost uniform proportion to the height of the elevation of the bed. Where the height of flood is nothing the elevation of bed is also nothing—at Damietta. In 1800 years, it is thus seen, that for every foot high of the flood at Cairo, the Nile has elevated its bed 2.80 inches, at Thebes 2.34 inches, and at the first Cataract has elevated its bed for every foot of flood, 2.70 inches. This furnishes for streams perfectly analagous in all particulars to the Nile, an approximate scale for estimating the rate at which they elevate their beds while undisturbed by Levees in the distribution of their detrital matter over the adjacent countries. But while such is the rate of bed elevation in unleveed streams, we have seen, as reasoned to above, that the rate of bed elevation must necessarily be much more rapid in rivers confined by Levees. But one special fact confirmatory of this general proposition is, however, within our reach. The Rhine, which is Leveed from the sea almost to its source, has since the Christian era elevated its bed at the City of Mayance, 13 feet 4 inches. The Levee influence in this case has been in operation for but 300 years; and, therefore, assuming the rate of elevation in the river when it overflowed its banks the same as that of the average of the Nile, the bed-elevation for the 1500 years of overflow must have been 6 feet, and for the 800 years of Levees be so much as 7 feet 4 inches, or six-fold as great. The flood-level of the river Po, it is true, is said to be higher than the roofs of the houses in the city of Ferrara; but this statement is so loose that it may mean very much or very little. If the houses referred to be but one story high, the flood-level described in the statement may not be higher above the streets than ten or twelve feet. In London to-day, it would not be considered wonderful if we heard that the Thames, during high tides, stood as high as the eaves of some of the small houses in Blackwall, south of the Thames. And in New Orleans it would not be at all surprising to learn, that during the late floods, the water of the Mississippi stood higher than the roofs of some of the little squat cottages on the edge of the swamp sloping toward Lake Ponchertrain. Originally, marine swamps, as London, New Orleans and Ferrara, had been, it is after all not so very remarkable that the levels of those swamps should be found now, as they doubtless have been from time immemorial, considerably depressed below flood-water. Seeing then that the record is so loose in the case of the Po, it may be assumed that while that record points to a great elevation in the river-surface since the construction of its Levees, such an elevation, from the manner in which it is stated, must not of necessity be held as by any means alarming. So much for the reasoning and the facts as to the elevation of river-levels, whether the rivers be or be not confined by Levees. This question of bed-elevation and, therefore, of surface-elevation, has been made a great bugbear in reference to the embankments along the Mississippi; but when the few facts known in the case are subjected to examination, only such planters as take a very active interest in their great grandchildren will, while reclaiming the magnificent wastes of the Mississippi, trouble themselves by the reflection that after the reclamation of those lands, they may revert in some future century back to swamp, on the ground that the works of reclamation tend to elevate the flood-level of the river, according to the experience of a city 300 miles up the Rhine, at the rate per year of less than one-third of an inch!

Incidental to the question of Levees, a few remarks may be added on the subject of Debouch-bars. In a Delta these bars mark the shallowest water of its respective passes; the volume deepening up-stream until, at the junction of the passes, it reaches its general depth. The Rhone, at Arles—20 miles from the sea—has a depth of 43 feet, whereas the depth of water on its bar is but 6 feet 6 inches. This river has five passes or mouths. The Po di Volano—one of the passes of the River Po—has a depth on the bar of but 2 feet 6 inches; while some seven miles up-stream, that depth increases to ten feet. The same general fact has been observed at all the seven passes of the Nile, and of the numerous passes of the Ganges. This law of Delta-debouch is illustrated forcibly in the case of the Mississippi. The South-west pass—the deepest of the whole, has, according to the United States Coast Surveys of 1851 and '52, a depth of about 13 feet, whereas, according to Sir C. Lyell, the river has a depth at New Orleans of 168 feet.

In Deltas, rivers always divide into branches. Consequent on this branching the loss of volume in each outlet results—by the great increase of friction, &c.—in a loss of momentum. This loss of momentum, lowering the aggregate carrying-power of the stream, results in a proportional acceleration of deposit; and therefore, going on from its starting point—the branching—under the effects of a constant retardation, reaches its limit on the pass-bar. This, then, is the point of greatest deposit, and therefore of least depth; whereas, the branching point is the point of least deposit, and therefore, of greatest depth. Thus we find the Rhone, the Po, the Nile, the Ganges, like the Mississippi, all shallow in their passes, and deep above the separation of those passes from the main channel. These facts and reasonings on Delta-bars point directly to the natural remedy for lowering the water-line through a Delta and removing its bars. The diffusion of the water-flow being the cause of those evils, their remedy lies clearly in its concentration. The condensing of the whole volume of a stream in one channel will, by increasing its momentum, give a carrying power that will remove and transport far out to sea, the silt that, with an inferior carrying power, sinks into the bed of half a dozen passes. The improvement of river-beds, whether for the purposes of navigation or drainage, ought never to lose sight of the prime importance of *concentrating the flow*, in order by thus increasing the momentum—the "scouring" power—of that flow to remove the greatest possible amount of deposit from the bed, and thereby deepen the channel ; to propel that deposit out into the distributing currents of the sea, and thereby retard or stop altogether, the extension of the Delta. This conclusion is confirmed by the experiments of Gennétte, the observations of Guglielmini, and all the subsequent experience of the most respectable practitioners in Hydraulic Engineering.

The bars of the Mississippi mouths are subjects of great importance to commerce. The report of attempts to remove one or more of those bars by dredging, is incredible. Such an effort were a repetition of the story of removing the soil of the Augean stable. The mechanical power engaged in piling up material across the passes of the Mississippi is that of the Mississippi itself; and it were the rankest of folly to attempt to undo the *constant* work of that power by the puny efforts of some 100-horse-power dredge. The Mississippi itself is the only power that can be brought to bear in the case to undo permanently the work of the Mississippi. The Clyde, a century ago, did not present a navigation-depth of over three feet as high as the City of Glasgow ; but, though the bars and general bed were hard gravel, such has been the effect of concentrating its waters between regular lines of wharfs and jetties that it, to-day, bears to the Quays of Glasgow sea-going vessels of some 20 feet draught. Concentration then of its waters in one channel is the only means for removing permanently the Mississippi bars; and thereby preserving for New Orleans a commerce that otherwise must become every day more embarrassed as the Delta-advance adds uncertainty, difficulty, and danger to its communication with the sea.

But the commercial ground applies also to the other grounds of this course. The concentration of the waters of the Mississippi will not only assist shipment by removing the pass-bar, but will assist drainage by keeping down the water-line. The greater the momentum at the mouth, the greater the power of the river in displacing sea-water, and the greater the displacement of sea-water, the greater the outflow of river-water. Thus then does the concentrating of the stream tend to the depression of the up-stream water-level. But the elevation of the water-level in Delta-rivers has been shown above to go on steadily with the extension of the Delta—a mile of extension in that of the Mississippi being taken to represent an elevation in the flood-level at New Orleans, of $1\frac{1}{2}$ inches, at Friar's Point of $5\frac{1}{3}$ inches, and of $6\frac{1}{4}$ inches at Cairo. While the increased displacement of sea-water, as suggested, leads to a proportional lowering of the flood-level, the full effect of that lowering will be experienced permanently by the removal of that constant cause of increased elevation—Delta extension. Now the increased momentum resulting from concentrated flow, in displacing an increased amount of sea-water, operates necessarily farther out at sea; and, in so operating, bears the material of river-flow more thoroughly within the distributing influence of the Ocean-currents. The Amazon with its single outlet rushes into the sea with a momentum that forces its earth-laden water out into the Atlantic Ocean for 300 miles. The sea left thus to dispose of the material brought down by that great river, the Amazon has, as a consequence, no Delta. Concentration of its waters will accomplish like results for the Mississippi; and indeed the Mississippi is much more favorably circumstanced for the accomplishment of those results, in consideration of the direction and position of its outflow in reference to that great distributing agency—the Gulf-stream. The availability of the Gulf-stream as a distributor for the Mississippi may be inferred from the words of Sir Charles Lyell: "that drift timber from the Mississippi is carried to the shores of Iceland and Europe, and that the fine sediment at the velocity of the Gulf-stream would reach the point of Florida before sinking, and what was not deposited there would even be carried much farther on." Concentration of the water then will not only improve navigation by removing the bar; but, by increasing the momentum, will, in the resulting increase of outflow, lower the water-line: and, in the resulting limitation of the Delta-growth, will also remove the resulting constant tendency to the elevation of that water-line.

Having glanced at the special question of the dredging of the Mississippi bar, it may be excusable for glancing now at another question of the same class—Cut-offs. The Levee being the special object of our consideration here, no other deviation from it shall be made than that which it is now purposed to enter on.

The circuitous character of the Mississippi and its tributaries is sometimes attempted to be remedied for the purposes of drainage, by opening across the narrow part of a bend-peninsula a direct channel. This direct channel is known, locally, as a "Cut-off." Now, the current being regulated by the *rate* of fall, and the *rate* of fall between any two points being regulated by the distance between those points, the shorter that distance the higher will be the *rate* of fall, and the more rapid will be the current. If the fall be four feet from the beginning to the end of a *twelve-mile-bend*, then is the rate of fall in that bend four inches in the mile; but, if that begin ning and that end be connected by a direct channel of *four miles* across the bend, then is the rate of fall increased to 12 inches per mile. The velocity, all things else being equal, increases directly as the fall; and hence does this increase of the rate of fall from 4 to 12 inches increase the velocity, all things else being equal, *three-fold*. But the momentum of the stream, all things else being equal, increases as the *square* of the velocity; and consequently, when the fall and velocity are increased 1½ times, the momentum is increased 2¼ times; when it is increased two-fold, the momentum is increased four-fold; and when, as in the case of the Cut-off supposed above, the fall and velocity are increased three-fold, the momentum is increased nine-fold. Immense accessions of mechanical effect are thus seen to be evolved by Cut-offs. Now, in ascending the Mississippi, a steamboat encountering a current of five miles an hour, expends in the encounter a mechanical effect of suppose 25; then will that same steamboat, in encountering a current of six miles, expend a mechanical effect of 36; in encountering a current of seven miles, expend a mechanical effect of 49; in encountering a current of eight miles, a mechanical effect of 64. Navigation-resistances running up thus rapidly for every increase of current—or shortening of channel—the point is soon reached by such shortening, where steam-power becomes totally absorbed. Thus then, do Cut-offs endanger the continuance of navigation. This abstract reasoning, very true, is disturbed by the practical facts. If the soil cut through were indeed strong enough to withstand the accelerated current, that acceleration would continue to act through a proportionally contracted cut; but after a while, the effect of this acceleration, in the constant tendency of the flow to adapt itself to the material of the banks, tells in the gradual widening of the new channel to something like the *general* section of the river. With ordinary sections thus obtained for itself, the full effect of shortenings on the increased rate of flow, consequent on increased rate of fall, can apply under, only the supposition of free-outflow at the lower end of the Cut-off, and accelerated supply at the upper end. The engorgement of the channel below and the exhaustion of the channel above, tend, it is true, to divide the effect of the Cut-off between an increase in the velocity within it, and a lowering of the water-line from its lower end to a point considerably up-stream. This modification of the fact of increased velocity, however, must not be held to obviate it altogether. Cut-offs, notwithstanding the corrective influence of channel widening, of engorgement below and of exhaustion above, *tend* by their rapid rate of acceleration in river-resistances to embarras, and under circumstances perfectly supposable, even to exclude navigation. Every impediment to navigation involves an addition to the cost of shipment; and hence do the planters who seek relief from a Cut-off, entail (until at least the river shall have restored its disturbed bank-current equilibrium) on all shippers up-stream a greater or a less increase of shipment-tax on their up-stream freights. A Cut-off, then, may thus not only put a whole country under contribution, but may actually deprive it altogether of the benefits of water-carriage.

But navigation is not the only interest involved in protesting against Cut-offs. Increased velocity introduced at any part of the river-channel, while the velocity below that part remains undisturbed *per se*, the result will be that the waters, deposited at the termination of the increased velocity more rapidly than they can be passed off by the receiving velocity, will, *as compared with their reduced level within the Cut-off*, be "ponded" up. True, the additional momentum received by the volume of less velocity, will increase that velocity until at some distance down stream the effect of that additional momentum shall have been exhausted. This fact does not destroy the fact of "ponding" up, but by reducing the "ponding" at the point of termination of the specially accelerated velocity pushes farther down stream

—to the point of exhaustion of the additional momentum of the special acceleration—the point of greatest " ponding." The up-stream result may now be glanced at. Passing off the water at a velocity more rapid than that at which it is received, the Cut-off after a while reduces the level of the water in the old-channel; and this reduction of level, accompanied under the " suction" of the Cut-off with an accelerated velocity, extends up-stream to a point at which the Cut-off " suction" ceases to act. The Cut-off then alters the water-level to a sort of concave curve, beginning up-stream and ending down-stream, the deepest depression being within the Cut-off itself. This curve extends along the whole length of increased velocity of the flow—that increase ending up-stream at the point where the "suction" of the depressed-level of the Cut-off ceases, and ending down-stream at the point where the accelerated momentum of the increased fall or velocity terminates. The Cut-off then, is undoubtedly servicable in lowering the water-line between those extreme points, the lowering in the Cut-off itself being greatest; nor is it open to the drawback charged upon it popularly of overflowing the country down-stream. The increased velocity of the Cut-off, being accompanied with a reduction of level, discharges no greater quantity of water in the same space of time than that discharged by the original velocity, and original volume. How, indeed, can the Cut-off be supposed to discharge more water than it receives, or to discharge water more rapidly than that water is received? It discharges only the quantity it receives; and receives only the quantity that time for time had been received and discharged by the original volume. The popular objection to an occasional Cut-off of flooding down-stream is seen thus to be unfounded. And here it may be observed that in considering the effects of Cut-offs on navigation as well as on discharge, the remarks made in each case have been confined to *occasional* Cut-offs. A system of Cut-offs carried up-stream to the supply points of the rain-basin would, however,—*until the river should have re-established its original regime*, or until the surcharged channel should have worked out those modifications of depth, by which rivers usually dispose of accumulated waters—present the question of discharge in another light; for the shortenings of 150 miles in the lower reaches of Red River point to a continuance of those shortenings to an extent that will cause the delivery of the flood-waters of that River in three or four-fold volume into the Mississippi. This occurring at periods of like delivery in the other tributaries of the Father of Waters—all discharging under the acceleration of Cut-offs—the result would, until at least the river should have adjusted its depth to its accumulated floods, threaten along the whole Delta of the Mississippi terrible inundations.

Grave objection rests also against Cut-offs in the extent and degree of their increase in the velocity of river-flow. There is as suggested already a sort of balance between the cohesive strength of a river bank and the abrasive energy of a river-current. When the current exerts on the bank an energy greater than the cohesive resistance of the bank, the result is expressed in caving, shoals, bars, and alterations of channel. The tendency of a river is to go on making changes in its course until the equilibrium between the strength of the bank and of the current are fixed; and this equilibrium is one of the prime objects of the river in endeavoring to establish its *regime*. In rocky channels, streams dash over cataracts; in beds of boulders and compact gravel they rush along in almost foaming rapids; whereas within alluvial banks they invariably sink down into a gliding flow. In the latter case the total fall may show a high rate of descent; but the result of disproportionate velocity over the soft soil has settled down, after running through since the dawn of creation the programme of bars, and shoals, and caves, and lakes, and new channels, and old channels, into the sinuosities of to-day. Nature in all this is working by rule—a rule that, however it may be modified, can in no case be safely broken. The Cut-off then is a direct interference with the constantly operating law that rivers are in eternal progress towards their *regime*. By disturbing the balance of flow-strength and bank-strength, as struggled to by centuries of natural operation, the Cut-off simply succeeds in throwing back the progress of final result on the part of the river into the early stages of the world. Nature at once sets about defining its laws in such cases; and hence do Cut-offs, in accelerating the energy of river-forces, endanger from end to end of their resulting increase of velocity, violent changes of bank and bed. No Cut-off then can for any time continue to be the bed of the Mississippi River while the soil of the Cut-off is mere soft alluvium. Cavings of the most formidable character must be the consequence; and extending from end to end of the increased velocity consequent on the Cut-off—receiving, however, their greatest development in the Cut-off itself—it is quite impossible to tell where they may begin or in what form of evil they may terminate.— On Red River they may result in the restoration of its ancient outlet to the Gulf; and thus flooding the whole of Western Louisiana, turn with the characteristic suddenness of a torrent into the Atchafalya. On the Lower Mississippi the Cut-off may reduce to a permanent swamp, either the valley of the Yazoo, of the Lower White River, of the Lower Arkansas, of the Lower Red, or by causing the diversion of the Channel into Manchac, may, in twenty-four hours of its flood-season, reduce the whole of Eastern Louisiana from a teeming plantation to a miserable Lagoon. The Cut-off then, while undoubtedly calculated to lower the adjoining flood-level for the moment, is highly dangerous to navigation, and still more highly dangerous, whether as an agent of accelerated aggregation of water, or of accelerated velocity of flow—to all the great interests of life and property on the rich alluvium of the Mississippi Delta.

CHAPTER III.

THE LEVEE.

WATER standing in a vessel or enclosure of any kind, presses with equal force on the bottom and the sides. At ten feet deep the pressure of a column of water of a foot square is the weight of that column—10 cubic feet at 62½ lbs. per cubic foot—625 lbs. The bottom of the vessel containing this water of ten feet deep, bears a load, therefore, of 625 lbs. to the square foot; and the sides of the vessel at the junction with the bottom, bear the same strain. The pressure of standing water, it will be seen from this explanation, increases as its depth; being for 20 feet deep, 1250 lbs. to the square foot; for 40 feet deep, 2500 lbs. to the square foot; and this pressure is for the same depth, precisely the same, square foot for square foot, at the sides as at the bottom. The *width* of water it will be seen from this proposition has no influence whatever on its *side*-pressure; the width of the Atlantic Ocean, exerting only the same hydrostatic pressure on the shores as a mere *thread* or film of water of the same depth. The popular opinion that the width of the Mississippi affects a proportional pressure on the Levee, it may be remarked here is an error. The side pressure, or in other words, the weight of the water-column at the several points of its depth, goes on increasing from the surface, where it is nothing, to the end of the first foot of depth, where it is 62½ lbs. to the square foot of side; to the end of the second foot, where it is 125 lbs. to the square foot of side; and so on, the side pressure at any depth being for every square foot of side, the product of 62½ lbs. multiplied by the number of feet in depth of the point at which it is required to find that side-pressure. Supposing them to be exempt from the blows of waves, and of currents, the pressure exerted on Levees would then be in the proportion, at every foot from its top, of 1, 2, 3, 4, 5—a regular arithmetic progression from 0 at the top, to the base—the section representing pressure being thus at the base, the same number of feet in width as the water is in depth. This gradation of pressure in standing water at its several depths, presents the following geometrical form:

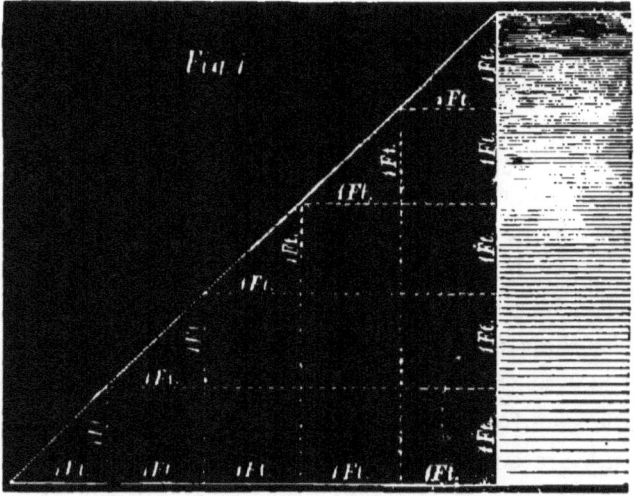

The pressure of the Mississippi then, on its banks—rejecting that from the blows of waves, or of currents—varies at the several depths as the widths vary in the above figure; and hence will that pressure be resisted effectively by any earthy matter, impervious to water, embanked in the above form.— *Any* earthy matter, provided it be impervious to water, piled up in the above form, will discharge the quiescent pressure of the water, because *all* earthy matter is *heavier* than water;

and consequently the above section if made of *water*, will not exert a pressure able to overcome the dead weight of the above section, if made of *earth* or other material heavier than water. But this reasoning rejects all other pressures of the river than its pressure as standing water. Great forces, however, at, especially, the first few feet in depth are exerted on Levees, over and above the standing pressure, by the incidents of waves and currents. A direct cross-wind, in a reach of a mile wide, discharges the dead-weight of the water upon the bank with the velocity of a wave ; and, therefore, occasions a great accession to the standing pressure of the water for the depth of that wave. The steamboat of the Mississippi, as another producer of wave-motion, is also an agent converting the *standing* pressure into a multiple of that pressure by velocity ; but the steamboat-wave, acting on the Levee obliquely, produces an increase of pressure proportionately less than that produced by the wave striking it under the impulsion of a wind blowing against the Levee directly. No practical measure of this particular cause of increased pressure on Levees is obtainable ; and, therefore, is this cause disposed of here without any attempt to estimate its measure, in the form of a specific quantity. The wave-blow, whether resulting from wind or boat, is a contingency of Leveeing that must be met, as involving an unavoidable necessity of increase on the size of Levees, over and above that necessary to balance the pressure of standing water. The other head of increased pressure, over and above the *standing* pressure on Levees, is that of current-blows. Previously to this the force of currents has been referred to in general terms ; but in order to express the importance of that force in the present case more fully, it may be well to present it here, in the form of a specific quantity. Mechanical effect is measured in the compound quantity of the weight moved, and the distance through which it is moved. "Feet-pounds" is the denominational term employed to express this effect. 100 Feet-pounds represents the mechanical effect expended in removing 100 lbs. one foot, or 10 lbs. ten feet, or 1 pound, one hundred feet—the mechanical effect expended being in each of these cases, the same in quantity. A current striking directly at, say 6 miles an hour, strikes with a velocity of $8\frac{1}{2}$ feet a second. This $8\frac{1}{2}$ feet multiplied by itself, (or squared) gives a product of $72\frac{1}{4}$, and this $72\frac{1}{4}$ divided by the *constant* quantity 64·4, shows a quotient of $1\frac{1}{8}$. The quotient so obtained is an abstract quantity, representing the multiple necessary to apply to the dead weight of the striking body in lbs., in order to bring it for a velocity of six miles an hour, to its mechanical equivalent in feet lbs. Suppose, now, that a current acts at a velocity of 6 miles an hour for a section of 20 feet deep, then the gross average pressure of this section on the bank, as for *standing* water, being 625 lbs. per foot, in length, the mechanical effect expended against the Levee will be, for every foot in length, 625 multiplied by $1\frac{1}{8}$, or 700 feet lbs. The mechanical effect that will move 700 lbs. one foot, will move 8,400 lbs. one inch; and there being little or no elasticity in a solid bank of earth, the current-blow that forces it back a *couple of inches*—repeated as that current-blow must be assumed to be—may be held sufficient to force it back *altogether ;* and therefore, finally, to sweep it away. Such then is the practical value in Leveeing of the force of currents. The wave-blow, as has been remarked, is an unavoidable contingency of Leveeing, but then it must be recollected that while some observers go to the extent of alleging that waves do not involve any increase whatever of pressure latterly, be that pressure what it may, it is at all events confined to the height of the wave—a height that in the extreme case on the Mississippi does probably not exceed 18 inches. The wave-blow, then, involves no very formidable accession to the strength of the Levees. Current-shocks, however, are of a very different character ; but on the other hand, unlike the wave-blow, are altogether, or to a great extent, avoidable. The force of a wave and the shock of a current represent the aggregate pressure that may be brought to bear under the most unfavorable circumstances, in swelling the dimensions of the Levee beyond the form required under the above reasoning, for the pressure of *quiescent* water.

The time will come when flood-waters will be excluded from the magnificent low-lands of the Mississippi, at the cost of hauling, from wherever it can be obtained most conveniently, the best material for embankments. The material at hand will continue to be used for some time ; and therefore does it become a matter of necessity to use it with a knowledge of its advantages and its disadvantages. Sand, loam, and clay are the materials at present employed for the construction of Levees, the loam and clay, unfortunately, in small quantities. The weight of water, it will be recollected, is $62\frac{1}{2}$ lbs. to the cubic foot ; whereas that of light sand is 95 lbs. to the cubic foot ; of loam 124 lbs. to the cubic foot ; and of stiff clay 135 lbs. With such a difference as that between 95 and 135 in the materials found at different points, it becomes a matter of importance in designing the cross-section of large Levees, to consider the specific gravity of the material to be used. A varying cross-section of Levee is consequently a necessity of a varying soil. The Commissioner, it may be observed here, was censured at the time by some parties for having given the Levee across the Yazoo Pass and Levees of that locality, a larger cross-section than that previously adopted as a rule of general application ; but that gentleman would have made a grave mistake, for which his own judgment and perhaps the popular judgment would have censured him to-day, if in determining the dimensions of those Levees, he had not gone to the full extent demanded by safety on the score of *weight*, in exceeding a standard that, *if* well adapted to the *average* material of Levees, was certainly ill adapted to the only material obtainable in the cases in question—principally light sand. But besides objections based on the gravity of the materials, others also apply, classifying them into different degrees of adaptation for Levees. Wash and percolation are two most powerful agents of destruction in the case of river-embankments; and hence, does it become of the gravest importance, where the choice can be made, to select such materials as are most cohesive and impervious. The lightness of a sand bank is but a small disqualification for Leveeing compared with its liability to wash and leak. Its "wash" is not even confined to wave, current and rain; but is carried on actively also by the wind. Sand is liable not only to run and blow away in a dry state; but in also a wet state is liable to run, or "melt" like so much sugar. But while its lightness lays it open as a material for Levees to great objection on the ground of duration, the worst of its properties in such works is its liability to percolation. A bank of ample section to resist the total pressure brought to bear on it, when that pressure acts from the *outside* slope against the whole weight of the bank, will yield when that pressure becomes transferred from the *outside* of the bank to some point or plane *within* it. In the latter case a portion only of the whole mass is engaged in the resistance of the whole pressure. Now percolation of the water into the body of the work, places the Levee under these very circumstances. A thread or plane of water, finding its way into the interior of an embankment, exerts just as much pressure against the earth on each side of it as if that thread or plane were an ocean of the same depth as that thread or plane. As this thread separates the parts of the Levee, the outside water fills up the split or open; and thus preserving the sand height of water within the split, as at the beginning of rupture, the Levee becomes completely rent asunder; and thus reduced in its aggregate power of resistance, is finally swept away. Porous materials then in water-banks, no matter what be their weight in the banks, tend by the insinuation of water threads between their parts, to destruction of those banks—this tendency, however, being greatest at the time of the construction of the works, and least at the time when their adhesion shall have been perfected by the coating by deposit over their external faces, and the insinuation by filtration in their internal pores, of earthy matter. Loam is much better for water banks than sand. Thirty per cent. heavier, it meets all the conditions involved in Leveeing on the ground of weight so much better than sand. Much stauncher in its parts, it is superior to sand in all those serious objections applying to sand for the purposes of *water-tight* embankments. The very best of those soils obtainable under the present practice on the Mississippi for the purpose of river-banks, is blue clay. Several kinds of this clay are found on the lines of the Levee-works; but they are all subject to the disadvantage of a greater or a less admixture of fine sand. Perfectly impervious to water as they all are, the presence of sand lowers their usefulness partly by involving a lighter weight, but mainly, and sometimes even to a very serious extent, by giving them a tendency, especially after frosts, to melt or run like marl in water. But notwithstanding these draw backs, *the clays* of the Mississippi bottom furnish its very best material for Leveeing.

The different bulks necessary with different soils for the same Levee, has already been pointed out as an item of consideration in the use of the materials entering at present into Mississippi embankments. The remarks under this head were, however, confined to the influence on the subject of the different specific gravity of those materials. Another consideration in the premises rests on the fact of differences between their "angles of friction," or in the differences between their natural standing angles or slopes. Experiments recorded in Engineering authors of high personal and professional standing, set the angle of repose, or standing angle, of sand at an angle of 30 degrees with the horizon; of firm loam of from 36 to 45 degrees with the horizon; of clays at 55 degrees with the horizon. Using a form more acceptable to the popular understanding, it may be explained that those experiments show the standing slopes of those materials to be as follows:

For Loam - - - - -	1 foot high to from 1¼ to 1 base.
For Sand - - - - -	1 foot high to 1¼ foot base.
For Clay - - - - -	1 foot to ⅞ foot base.

Experiments of this sort cannot be disregarded; and therefore, though these figures *do* seem to savor rather more of the closet than of the field in the rapidity of those angles of repose, they are not to be discarded in any reasonings to the practical execution of earthworks. Coupling then the different specific gravities of sand, loam, and clay, with their different angles of repose, an assimilation of the merits of the three—waving the question of wash and percolation—may be made in terms of the limits of haul, at which it ceases to be economic to reject sand on the spot for loam and clay in the distance. In order to reduce the loosening and lifting of the earths to a common standard, let it be assumed that what might be saved under that head in sand as compared with the other two, and with loam as compared with clay, are balanced by the superiority of clay over both the others, and of loam over sand, in weight, strength, and imperviousness. In consideration of the vegetable matter permeating loam, the porosity permeating sand, and the lightness and friableness of both, the advantage possessed in these respects by clay are hardly overstrained by being set down for the present purpose, as fully equal to the advantages possessed by sand and loam over clay, as a material for "borrowing-pits." The retentive properties of clay, and in a less degree of loam, may be said to increase the difficulty of excavations in that material in a flat country; but the clay of the Mississippi flats, resting invariably in thin layers *on sand*, the shallow and wide-cuts necessary therefore, for clay-pits, may be made perfectly dry by running up through them at *starting*, a narrow tap-drain *to the depth of the sand.* Assuming then the three materials equally costly for loosening and lifting, this equation of their merits may be examined as a question of haul. The following figure shows in the broken line, a cross-section of Levee having a crown of 3 feet wide and a base of 21 feet—the side-slopes corresponding to the angle of repose for sand, that is to say, cor-

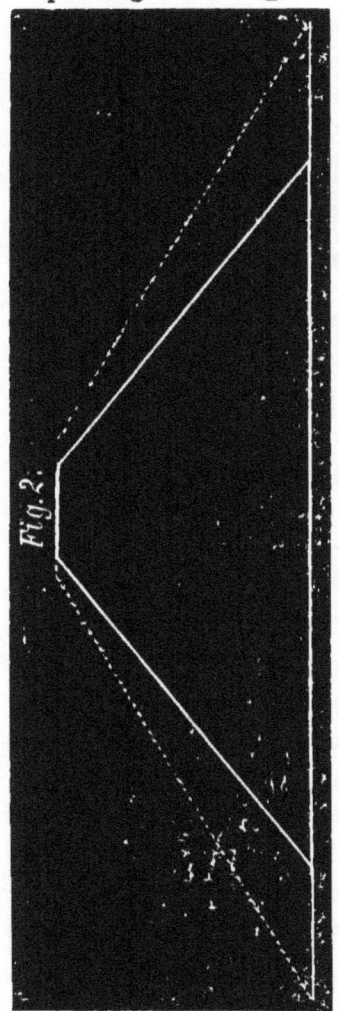

responding to the least angle at which sand will stand. The white line in the figure shows a Levee of clay, the crown being 2 feet and the base 15 feet—every part of this latter being deduced from the former in proportion of the weight of sand to that of clay—95 to 135—and therefore presenting, at all points, a resistance to the horizontal thrust of the water equal to that presented at corresponding points in the larger section (the broken line) of the Levee of sand. Two feet wide at crown, for example, presents as great a resistance in the case of clay, as 3 feet wide does in the case of sand; 15 feet wide at base presenting as great a resistance with the use of clay, as 21 feet wide at base does with the use of sand. This figure, then, illustrates the effect of difference in the weight of the two materials in regulating the size of Levees. But this difference is still further increased under considerations arising out of their varying angles of repose.—The increase of pressure has been already shown to go on from the top, under extreme conditions at an arithmetical progression; and this arithmetical progression of Levee-pressures or strengths has also been shown to express itself practically in an aggregate side-slope of 45 degrees—a total slope of one foot base for each foot in height. The preservation of equal strength at all parts of the Levee does not require, therefore, a greater width under even the most unfavorable circumstances than (whatever may be the proper width of crown,) side slopes from *each* side of crown at a rate of one-half foot horizontal to one foot vertical. Such a section may be said to be, in general, the section of uniform strength. The strength of any thing being, according to the mechanical axiom, the strength of its weakest part, an excess of strength at any one part is, it is almost needless to observe, a waste of material in Leveeing, and consequently a waste of money. In practice, however, it is impossible to conform to the section of uniform strength in Levees; seeing that the *controlling* consideration rests in the standing angle of material. The standing angle of clay has been set down at eight inches base, to one foot in height; and, therefore, may be held to conform closely to the section of perfect economy of material—the section of uniform strength. 21 inches of base for every 12 inches of height, being the standing slope of sand, that material is seen in the excess of its *natural* section over the section of equality of strength, to involve in Leveeing a very large waste of material; and, therefore, of money. In a Levee having a 3 feet crown, a 21 feet base, and a height of 5.20, as shown by the broken line in figure 2, the area of cross-section is 62.40 square feet. This Levee, it must be recollected, is one of unequal strength; and, therefore, measuring its effective strength by its weakest part—its 3 feet crown—we find the limit of its actual resistance to be, when made of sand, as (3 feet × 95 lbs.) 285. A clay Levee of 2.11 feet crown sloped down, at the standing angle of clay, to a base of 9 feet for 5.20 feet high contains *within* it the slope of uniform strength; and consequently, its crown being its weakest part, the limit of its effective resistance is as (2.11 X 135) 284.9. This clay Levee of 2 feet crown, and 9 feet base presents, then, precisely the same resistance to water-pressure as does the same Levee of the same height, having a crown of 3 feet, and a base of 21 feet. The cross-section of the clay-bank in this case, is 29 square feet; while, as has been said above, that of the sand is 62 square feet. But practice goes still further in increasing this disproportion between the different quantities necessary in Levees of sand, and in corresponding Levees of clay. The standing angle as presented in theory, must be deviated from in both sand and clay to meet, in practice, the contingencies of floods and rains. Lighter, looser, and less adhesive than clay, the flattening of slopes in sand below that of the angle, or slope of repose, must be much more considerable in practice than that in the heavy concreted and adhesive bank of clay, to resist, without endangering the effective strength or stability of the bank, the active washes of rains and waves. The practice, however, in these cases is so loose and various, that it cannot be expressed safely by a rule. Disregarding it altogether, however, and confining the equation of the two materials to the simple fact of the difference between their strict standing angles, 26 yards of clay are seen to be equal in a Levee of 5 feet high, to 58 yards of sand in accomplishing the object of all Levees—effective resistance to floods. In a Levee of 10 feet high, 43 yards of clay are as effective as 102 yards of sand; and in a Levee of 15 feet high, 60 yards of clay accomplish all the purposes of 146 yards of sand. At 15 cents per cubic yard, the difference in money between the employment in a 10 feet Levee of 43 yards of clay, and the corresponding quantity of sand, is $8.85 in favor of the employment of clay. Supposing, under this view of the case, the sand to be found on the site of the Levee, while the clay cannot be obtained without haulage from some distance, the great objection to the employment of sand for this purpose, suggests the enquiry: to what distance are the parties interested bound by the foregoing considerations to haul clay to their Levees? $8.85 are available, it has been shown under the foregoing comparison, for expenditure in obtaining the clay as compared with the cost of the sand on the spot; and this $8.85 distributed over 43 yards of clay, shows an available amount for the haulage of clay of 20 cents per cubic yard of clay. In numerous instances this rate per yard will cover all the inconveniences of haulage to the Levees, for a distance of half a mile. To sum up these remarks: it may be concluded that, supposing sand to be, under any necessity whatever, fit material for the construction of river embankments, taking the cross-section of equally strong Levees of the two materials, comparing those cross-sections according to the data furnished in the standing angles of the materials, and setting down for the moment, the lightness and porosity of sand, as compared with the heaviness and imperviousness of clay as material for water-banks, at the mere difference in cost of excavating the two, all the immense advantages of clay over sand, not covered by the assumption made here in the case, may be secured at the same cost as sand, by hauling clay to the site of a Levee from a distance of half a mile. In practice, this undoubted fact and the imperative duty that it points to, may be found one of very frequent and profitable application; for, in several cases, clay can be found in abundance in the Yazoo Valley, within half a mile of existing Levees of almost unmixed sand. In the bottoms and banks of the creeks of the out-fall behind the Levees, in the beds of the old and dry lakes, so common behind those Levees, in the hundreds of cypress-swamps in the neighborhood of the works, and on the surface of the higher lands, the contractor will find large quantities of strong, pure clay. The make-shift river-embankments of the State of Mississippi, before those works took the shape of a system, under the strong and able mind of Col. Alcorn, have already advanced to something like scientific conditions in plan and section ; but must make still further progress in order to fulfill all the conditions of cheapness and *efficiency*. Progress in those works points to a discrimination in the use of the materials; and, therefore, to considerations beyond those of the mere accident of the soil on which they are to be built. Sand is, in fact, utterly unsafe in a water-bank, and, therefore, unfit for any works designed for the protection of property from overflow. Break after break, in such Levees, is going on with its lesson of instruction to the necessity that first felt obliged to employ sand; and as the property suffering from such breaks, becomes more and more valuable, the time is approaching when the question of material in the Mississippi Levees will be considered by the owners of property behind them, as a question of *insurance*. Sand will, by and bye, be either altogether rejected in Levee making, or used only in positions where its properties can be turned to usefulness ;* and in order to open the way for this purpose in the right direction, the general question of its merits as compared with clay is here considered for the information of the planter. The unthinking will, probably, undervalue the reasonings employed in the case, as what a certain gentleman in Mississippi would call " College" nonsense ; but men of reflection will recollect that progress knows no road but that pointed out by observation, reflection, *and* calculation. Discrimination between the materials at hand is the first object aimed at in the foregoing remarks on those materials; the haulage of the best material for a short distance is the next object: and, following that, the haulage of the material to the full extent —as between sand and clay—of half a mile ; the next step in advance being one that is also yet to come—the total rejection of sand, as a building material, from the Levees of the Mississippi.

* See note, page 80.

The materials at hand will, however, continue to be used for some time in the Mississippi embankments. It becomes, therefore, important to consider the best means for reducing the disadvantages of their use to the smallest possible amount. The leaky property of sand is its greatest objection; but this may be overcome to a large extent by constructing within the bank a vertical wall from crown to base, of clay, thoroughly tramped and puddled. This "puddle wall" ought to contain no vegetable matter, such as grass or roots of any kind; and when wet to a proper consistency ought to be shoveled into the place left for it in the sand-bank as the bank goes up, layer after layer. When the puddle is in its place it ought to be tramped down well; it is indeed beaten down in water-banks in England and Ireland with a heavy maul or rammer.

The practice of cutting out a trench for the puddle, or "muck ditch" as it is called on the Mississippi, in the natural surface of the ground, is generally useless, and sometimes positively mischievous. Where retentive subsoils exist under the base of the proposed bank, then it is certainly a clear gain in staunchness to run down the puddle-wall of the Levee to a bond with the underlying impervious earth. But the experience along the shores of the Mississippi leads to the presumption that, in those cases where the sand does not commence on the surface, a ditch of three feet deep is more likely to present a bottom of sand than of loam, or clay. The rationale of those "muck ditches," as they are called locally, rests on their usefulness in preventing leakage; and, therefore, supposing the ditch and wall carried up regularly with a puddle, those ditches in a great majority of cases failing to reach a more retentive soil than that at the surface of the ground, involve in all those cases an utterly resultless waste of money. Besides, to undertake to prevent leakage through the porous earths of the natural shores of the river, is a hopeless labor; and so far as the strength and durability of the Levees are concerned is a labor, also, perfectly useless. It accomplishes nothing whatever, for the artificial embankment. But in some cases these "muck ditches" are, as already stated, mischievous. Across those lagoons or creeks which are dry during periods of low-water, the foundation for banks consists generally of a hard crust of clay for a few feet thick, overlying quicksands or thin puddles. These crusts, like the grillage of timbers used for the foundations of some Engineering works, are highly valuable in those situations, by diffusing the weight of the superincumbent Levee over a wide bearing ; and thus, though unequally loaded by the necessary cross-section of the Levee, assist, in proportion to their strength, to distribute that bearing equally. This, where not sufficient to obviate the sinkage altogether, reduces it considerably ; and in bringing a large area to act in the resistance, assists to guarantee with the least possible "sinkage," and, therefore, least possible loss of work—of money —a finally well-sustained foundation. The "muck ditch," however, cuts this natural platform for the Levee in two parts ; and *over* this cut, the greatest weight—that at the crown— pressing vertically, acts as with a leverage in bending down, and finally breaking off the natural crust of the surface. The necessity therefore follows, under those circumstances, of employing an excess of earth in forcing out laterally, and forcing down vertically, the running sand or soft puddle of the underlying foundation in order to compress those soft materials into a compactness sufficient to present an effective resistance to the weight of the superincumbent embankment.

Rejecting then the practice of cutting a muck ditch along the base of the Levee, it is recommended here that the earth of the base be *loosened* for six or twelve inches in order to secure, between the artificial and the natural bank, a proper bond. Indeed where the natural surface is loam or clay for any considerable depth, it would be highly judicious in order to prevent grasses or other vegetable matter of retarding the bond between the Levee and the ground, to skim the surface; and one good purpose thus served, the sod so skimmed off the base may, on the completion of the Levee, realize the further good purpose of staunching it by laying them in a close coating on its water-slope. Be the substitute, however, for the muck ditch what it may, the present practice in the case, if even not useless and unsafe, is certainly absurd when it is recollected that it is in fact a "muck ditch" with the important exception of the "muck."

The lightness of sand is a great objection against the use of that material in water-embankments. Sand is constantly carried away in immense quantities by the current of the Mississippi; and therefore, to invest money in banks of sand for the exclusion of the Mississippi floods, does not seem to be a policy very remarkable for its astuteness. As its porosity presents the use of sand in Levees in the shape of a question of safety, its lightness presents its employment in those works in the shape of a question of maintenance. Rains constantly washing the particles in its crown down to its sides; and washing those in its sides out upon its base, the weakest part of a Levee—the crown—is undergoing constant reduction in its dimensions, and consequently in its strength. Current-washes and wave-washes on the water-side co-operate in times of high-water with the rain-wash at other times, in reducing the strength of Levees thrown up in sand. Maintenance becomes thus in the case of sand Levees, a serious outlay. To remedy this—and indeed at the same time assist its want of imperviousness—the most convenient course is, to cover the water-slope of such banks with as thick a layer as may be obtained of clay, if obtainable, or if not with as thick a layer as may be laid on according to the above suggestion, from the quantity of earth obtainable by collecting the loam of the adjoining surface. In Ireland it is very common to face water-slopes with grass-sods laid on their flat beds with regular headers and stretchers, as in Ashlar work, the whole being cut down to the plane of the slopes. Under the direction of Mr. M. B. Hewson, I have myself conducted large quantities of this work for the Board of Public Works, under the measures for the drainage and navigation of Irish Rivers. In the South "sods" are not generally obtainable; but all the advantages resulting from their use in water-banks, may be obtained there by sowing the seeds of some southern grasses in a coat of loam-dressing on the slopes of those banks. Bermuda grass is well adapted for the preservation of artificial banks; but though often employed for that purpose on Railroads in the Northern States, is excluded from use on banks in the South-west by what would seem to be no better than a mere prejudice. The rapidity of its growth is not the only recommendation for the employment of Bermuda grass on Levees; for it possesses the further recommendation of growing in both shade and sun. During high-water it will catch a great quantity of sediment; and by the consequent annual coating of impervious earth, will add to the strength and durability of the Levee. The decay of the tops and blades of this grass will also assist in covering the water-slope of the Levee with an annual coating of impervious matter, and in the case of Levees built of sand, will thus tend greatly to the correction of their two great shortcomings—washing and leakage. A hedge of Osage Orange, set closely along the inside slope, by excluding both travelers and cattle, and a coat of Bermuda grass set on both sides, by obviating wash whether of rain or current, will save the parties interested in Levees, a large annual sum for their maintenance.

In Europe, generally, it is usual to protect embankments by growing on their top and sides thickly growing grasses. In parts of Holland straw is used for the same purpose. Twisted into ropes about 2 inches in diameter, it is laid on the face of the bank, and pinned down with hooked or forked sticks; rope after rope being added each in close contact to the previous one is so laid down until the whole slope is covered with a complete mat of straw. Vegetation in course of time commences underneath the straw, and blades of grass making their way between the ropes, the whole becomes a compact sheeting. Fascines, hurdles, and brush-wood, are sometimes employed for the same purpose. Large stone slabs are often used by Engineers in Bombay for the protection of the slopes of heavy embankments from the weather. So important is it found in experience all over the world, when it is worth while to go to expense in the construction of embankments, to go to further expense for their efficiency and preservation. Any thing that is worth being done *in water-works* is worth being done properly and well.

The standing slopes that have been given above for sand loam and clay are the standing slopes of those materials when dry. The dry slope and the wet slope of all earths are, however, more or less different. The same earth that stands in practice at an angle of 2 to 1 in a dry position, will require in a wet situation a slope of 3 to 1. Some earths have been found to require in a wet situation slopes so low as 4 to 1. A river embankment involves both the two distinct conditions of dry and wet slopes—the inside being necessarily regulated by the conditions of *dry* slopes, the outside being subject to the tests applicable to *wet* slopes. Wetness and dryness, however, do not cover the whole difference between the circumstances of the dry and the wet slopes of the Levee; for the outside slope, in addition to the disadvantage of wetness, is also subject to the further disadvantage of waves and currents. The practical facts of the case establish, therefore, the general proposition that the external or wet slope of the Levee ought to be less in rate than that on the inside, or dry slope. If the wet slope be sufficient for the necessities of its position; then, to carry out the dry slope at the same rate is a simple waste of material, and consequently, a waste of money. In six cases of well known water-banks in England, the inside or dry slopes vary according to the material, from an incline of 1 to 1 to an incline of 3 to 1, averaging an incline of $1\frac{3}{4}$ to 1, or about 30 degrees with the horizon; the outside or water-slopes, in those six cases, varying from an incline of $2\frac{1}{2}$ to 1, to an incline of 5 to 1—the six showing for the water-slopes an average incline of about $3\frac{1}{2}$ to 1, or about 16 degrees with the horizon. Natural water-slopes, formed under water, such as those of bars in the Mississippi, or other rivers, vary from an incline with the horizon of from 5 to 30 degrees—the average of these two extremes being $17\frac{1}{2}$ degrees, or nearly 3 feet of base for every foot in height.

The height of a Levee above high-water mark has been set down, by practice along the Mississippi, at $2\frac{1}{2}$ feet, and at 3 feet. A Levee system, as has been shown in another chapter, does not occasion an immediate elevation of the previously established flood-level. General considerations, then, have nothing to do with this head of the subject, seeing that it is a head proper to local specialities. The width of a river and the force of the winds regulate the height of its wind-waves; while the width and current of the river, coupled with the speed, load-line, and midship section of its steamboats, regulate the height of its steamboat-waves. Two feet would, probably, plumb the highest wave resulting from the accidental combination of the greatest wind-wave with the greatest steamboat-wave rising on the Levees of the Mississippi. The looseness of the observations made as to high-water in that river, coupled with the further consideration that those observations may not go sufficiently far back, in time, to embrace that particular combination of circumstances which produce the highest possible flood, suggest the propriety of basing the height of the Levees on a margin over and above the strict inch of the recorded high-water mark. Allowing 24 inches for the height of the maximum wave striking the Levee, 12 inches additional is certainly not too large an allowance for the contingencies of the case, in determining the height above high-water of the Mississippi Levees. Close enough, already, no reason certainly appears to show why, the standard of an excess of 3 feet above the highest known flood should be lowered; and it is, therefore, safer to conclude that experience, as in the case of the State of Mississippi, has settled the question between economy and safety in the matter, by fixing the height of Levees at three feet above the highest level of quiescent flood-water in the river. Four feet of an excess would, of course, be still safer.

The crown-width of Levees is a question less of rule than expediency. In England, water-banks have an average width at top of 3 feet. In Ireland, the top-widths of embankments for drainage, are about the same. In Holland, however, the Dikes, when employed for road-ways, are exceptionally wide across the crown. The Sea-banks of Holland are, in any event, no guide in fixing on the dimensions of river-embankments; nor, indeed, are the size of water-banks in England, or Ireland, quite a safe guide for such banks when subject to the wash of the immensely heavier rain-fall of the Lower Mississippi.— Local experience, then, is the best, and indeed, the only guide in this matter. In Arkansas, it is true the local experience has been had under circumstances which make it start from too high a point; with the view of a necessity for adapting the section of the Levee to the *width* of the river, some absurd and ignorant theory has led to the rule, that all Levees in that State be as many feet in width at crown as they are in height. These works, however, have been carried out, chiefly, without the guidance of professional skill. In the State of Mississippi, the practice has settled down into a width of 5 feet for the crown of Levees generally. My own practice and experience lead me, however, to the conclusion that 3 feet is sufficient to cover all the contingencies of rain-washings, cattle-tramping, &c., during, and for a sufficient time subsequent to, the hardening and cementing of the works. Economy, however, in this case, as in that of height and all other dimensions, is the only limit; for safety is always the gainer by an excess of section.

Shrinkage had been included in the considerations regulating cross-sections in the Mississippi practice. The generalizing pursued in this case was as erroneous in execution as in that of the " muck ditch." Different materials shrink in banks differently. Its particles—fine and loose—sand, however loosely it may be shovelled together, fills its space closely; and, therefore, whether wet or dry, *settles* at a very small diminution of its original bulk. In *time*, too, the process of this settlement is short. One-tenth of its original content is a liberal allowance for shrinkage in sand. Tough clay, however, is banked up under different conditions. Adhesive in its character, it is loosened and lifted in lumps; and from the size of those lumps, their shape, and their resistance to a change of form, they fall together in an embankment without compactness. Settling of such a bank is the process of filling up all the cavities and spaces existing thus between its parts; and hence in the case of lumps so large and stiff as those of clay excavations, is the amount of this settling quite considerable—generally about one-fifth of its original bulk. The time expended in the settlement of clay is longer than that in the settlement of sand. The different modifications of material between sand and clay settle as to time and quantity, in proportion to the respective amounts in their constituent parts of sand and of clay. An average of 16⅔ per cent. then, the allowance generally made in Mississippi, waving the objection to the principle in the case, is not, in all probability, a great error on either side from the strict justice as to the *quantity*. Two inches to the foot for the height, with proportional increase to the side slopes without any addition to the width at base, is added in practice to the intended settlement section of the Levee, in order to cover the loss of form and size by " shrinkage " or settlement. According to the variations in the height of the bank from the surface of the ground an addition of one-sixth was thus added on to it, over and above the gradient line of 3 feet above flood.

The section of a water-bank is a mixed question of theory and practice. In examining its several parts here, it has consequently been found unavoidable to mix up the abstractions of the subject with its working facts. Having, however, made those examinations under the several heads of slope, height, and crown, the next point to be made is the combination of the results in the elimination of the practical cross-section. And first for sand. The width of crown being taken uniformly at 3 feet, the slopes of a Levee, showing the strict angles of its standing slope, on the inside for dry sand and on the outside for wet sand, is represented by the light line in fig. 3. The broken line represents the section adopted in the present Levee practice of the State of Mississippi; the heavy line in that figure representing the section resulting from the employment of the surplus material used under that practice, in increasing the resistance of the material when distributed at the strict standing slopes of the wet and dry sides of the section, in a manner to produce an equal *increase* of the resistance of the material of those slopes to motion from wind or wash. The section, figure 3, applies to sand. It adopts the *quantities* of the practice at present pursued in Mississippi; and redistributes those quantities on the basis of the respective angles of repose of the material in dry and in wet slopes. Fig. 4 represents a cross-section of equal strength with that of fig. 3, the one being assumed to be carried out in sand, the other in clay. The strength of the sand-section being assumed for its basis, this clay-section is simply an addition to the wet and the dry slopes of repose of material equal in quantity to the additions made in fig. 3, to those slopes for sand.

The sections in the two figures given here show again the relative proportions necessary for equal strength in clay and in

EMBANKING LANDS FROM RIVER-FLOODS. 75

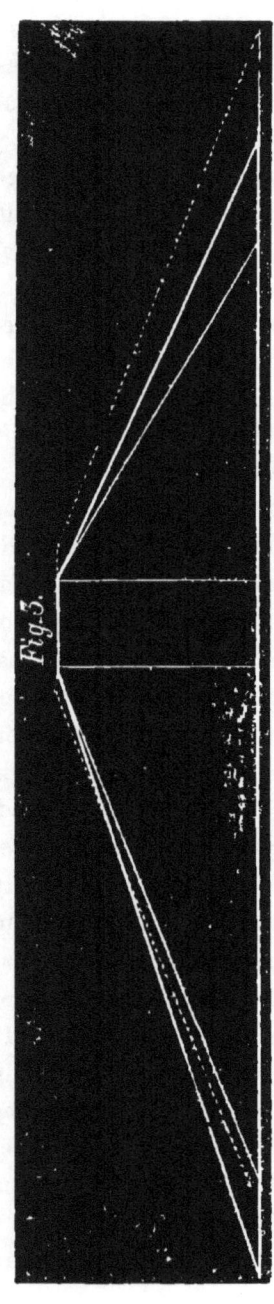

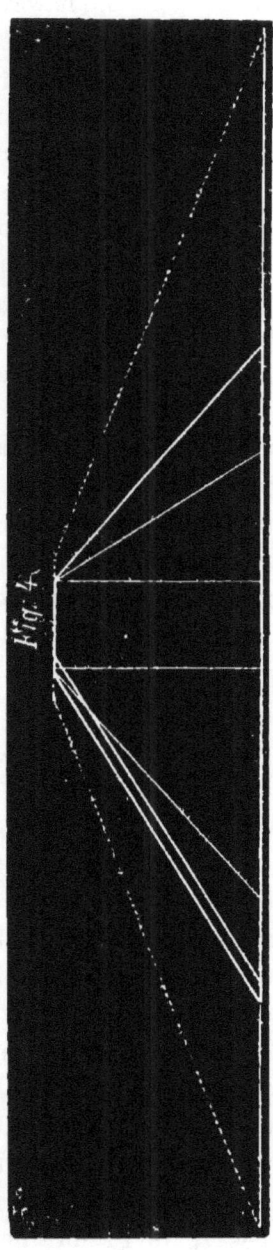

sand. The reduction, however, of the width of crown from 3 to 2 feet in the case of the clay is objectionable in practice. The considerations presented when reviewing the practice pursued in this particular in Mississippi suggest the expediency of adopting the width of crown in all cases, of 3 feet; and therefore, is the section above given on the supposition on either the inside or the outside slope of a Levee, of an addition of clay having a uniform thickness of 12 inches as represented by the second heavy line. This, then, shows a Levee as compared with that of the section in sand, of considerably increased strength—the excess at the point of least resistance—the crown—being over and above the sand-bank nearly 50 per cent. The standing angle of each material being taken as the basis of the respective sections, the additions made to those standing angles for the purpose of increased stability—the addition made in the case of clay giving the same additional width of base as in the case of sand—constitute in consideration of the superior weight and adhesiveness of the material, a greater additional stability over and above that of the section of the strict slopes of repose. The resistance to weather and current are greater, therefore, in the case of a clay Levee of the section shown in fig. 3, than of the section shown in fig. 4, thrown up in sand. The rule adopted in Mississippi for the proportioning of Levees is seen to be wrong in its recognition of equality of their dry and their wet slopes. But this rule is exceedingly inconvenient. Six to one being the proportion regulating the width of base in terms of the height in the State of Mississippi, the base for a Levee of 3 feet is 18 feet, of 6 feet is 36 feet, and of 12 feet is 72 feet. The crown in each of these three Levees being 5 feet wide, the base of the *slopes* themselves (deducting the width of crown) is for the 3 feet Levee 13 feet; for the 6 feet Levee 31 feet; for the 12 feet Levee 67 feet—the rate of slope on each side being thus: for the 3 feet Levee $2\frac{1}{6}$ to 1; for the 6 feet Levee $2\frac{11}{12}$ to 1; for the 12 feet Levee $2\frac{19}{24}$ to 1. Every height of Levee is thus seen, under the application of the Mississippi practice in this particular, to present its own peculiar slope; and consequently does the whole line of Levee in that state present a constant succession of *varying* slopes—a different slope for every different height. The inequality of the stability resulting from these circumstances is a mere theoretical consideration too trifling to be regarded seriously in practice. The objectionable feature of the case, however, applies to its practical inconvenience to the Engineer in estimating the quantities of the work. At this moment it does not appear that calculations *can* be made with *mathematical* exactness in such a case by any established formula; but be that as it may, it is very clear that such calculations must of necessity be tedious and complicated. The subject of calculation is, however, treated more fully in its proper place. The slopes ought to be laid down in terms of the height *exclusive* of the width of crown—a quantity that is a *constant* for all widths of base. To apply the inferences from the sections given in figures 3 and 4 to meet this expediency of equality of rate of slope for all heights, it may be observed that those sections are given for a height of seven feet. They show for that height a wet slope, for sand of $3\frac{1}{1}$ to 1, for clay of $1\frac{3}{4}$ to 1; and a dry slope for sand of $2\frac{1}{4}$ to 1, for clay of $1\frac{1}{4}$ to 1. For a height of 10 feet these slopes would for sand under the Mississippi practice be still flatter. As a rule then of constant application for all heights, these sections may be generalized into the following: in clay the inside slope to be $1\frac{1}{4}$ to 1, the outside to be $1\frac{3}{4}$ to 1; while in sand, the inside slope is $2\frac{1}{4}$ to 1, the inside $3\frac{1}{4}$ to 1. In consideration, however, of the fineness of the sand available along the Mississippi, and of the greater or less mixture of that sand in all its clays, and also in consideration of the necessity of simplifying calculations to the level of the expertness superintending the works, it is recommended here that the rule for constructing Levees on the Mississippi be (as the conclusion of the foregoing remarks on base, crown and material) as follows:

In Sand.

Inside or dry slope 2½ to 1.
Outside or wet slope 3½ to 1. } crown 3 feet

In Clay.

Inside or dry slope 1½ to 1.
Outside or wet slope 2 to 1. } crown 3 feet.

The clay-section supposes the rejection from the bank of all sand. For any *admixtures* of the two, no matter how small the proportion of that material in the admixture, it would not be safe to deviate from the proportions recommended above for sand. In order to show the ease with which this new practice may be substituted for the present faulty one, the following form of calculation is presented here:

Station.	Height of Levee.	Inside slope. 2½ to 1.	Outside slope. 3½ to 1.	Crown width.	Total width of base.
1	6.40	16.00	22.40	3.00	41.40
2	7.20	18.00	25.20	3.00	46.20
3	6.80	17.00	23.80	3.00	43.80
4	9.30	23.25	32.55	3.00	58.80
5	11.70	29.25	40.95	3.00	73.20
6	13.00	32.50	45.50	3.00	81.00
7	10.80	27.00	37.80	3.00	67.80
8	7.10	17.75	24.85	3.00	45.60

To guard against mistakes in making this calculation, it is recommended that after copying from the field notes the heights corresponding to each station, each of the other columns be carried out *separately*. Otherwise the multiple of 2½ in the one case will be often used by mistake for the other multiple and *vice versa*.

CHAPTER IV.

DETAILS OF LEVEE-WORKS.

EXPERIENCE is made up in Leveeing, as in all other works, of a knowledge of its details. Success in Leveeing, as in all other matter of practice, is regulated by paying due regard to small particulars. The intention of this contribution to the systematizing of those works excludes from this book a full examination of all the specialities that have arisen in the course of my experience on Levees in both Mississippi and Arkansas. Particulars, such as occur often or involve important considerations, are perhaps not excluded by the general plan laid down for my guidance in these pages. Attention may, therefore, be directed to a few considerations suggesting themselves by the special experience of my Levee-works. In preparing the ground for Levee-base it is necessary to clear and grub the whole thoroughly, leaving neither stump, root, brush, weed, nor even grass. This important duty is generally done with great carelessness. Before the work of embankment is commenced, all the timber, roots, weeds, and grass removed from the foundation ought to be disposed of in piles and burned to ashes. This rule should be enforced rigidly. It is the only means of guaranteeing the exclusion of all unfit material from the body of the Levee. In Arkansas particularly, and in Mississippi to a very large extent, to place logs, brush, and even whole trees, in the body of a Levee was an impropriety of not exceptional but of common occurrence. * In new Levees such an imposition can always be detected after rains by vertical holes in the crown and sides; and in dry weather may be detected by piercing the Levee at intervals along the crown with an iron rod. The only certain means, however, of excluding from the Mississippi embankments the materials grubbed and cleared from their base, is the enforcement of the rule that those materials shall have been burned in the presence of the superintendent before the bank shall be commenced.

In connection with this subject it may be observed here, that the clearing along the line of Levee ought to extend to all trees growing within their own respective lengths on either side of the crown of the Levee. All trees without that distance ought to be cut down; but if this should be supposed a need-

* The Coahoma Commissioner—who has made himself thoroughly conversant with the scientific principles and practical facts of Levees—in bearing testimony to the triumphant success during the late extraordinary flood of Levees properly planned and executed, calls attention to the cause of breaks being Logs, &c., in the bank. In his printed Address, of the 25th last July, that gentleman holds the following language:—" The question, then, is: did the Levee *when properly built* perform the work for which it was declared competent? I say it did; and challenge any man upon the whole line of Levee, on either side, *to point to a single break* where there did not exist *a local* cause." The address goes on to say:—" But in the large majority of cases where breaks occurred, the water either ran over the top, or *stumps and logs embedded in the work* occasioned the break."

One of the most general causes, however, of Levee-breaks during the late floods, has been the Craw-fish. This animal digs a hole through the Levee, from the water side, in order to obtain a passage through—the small fish, or water insects passing in with the flow, furnishing the object and the reward of his labor as prey. In sand, the Craw-fish cannot carry out his purpose, for the hole when made falls in, the Craw-fish, accordingly, desisting in his work. Clay banks are well adapted for the operations of the Craw-fish, and though, essentially the best in all other particulars for Levees, are open to this grave objection; this fact suggests that to obtain the general advantages of clay embankments in Leveeing, it is expedient in order to guard against its special disadvantages in those works, to carry up within them a wall of sand. The experience of the late flood makes this sand-wall in clay Levees a detail of the first importance.

less precaution, it certainly is not so in the case of all such trees *leaning* in the direction of the Levee. This should be done during the clearing of the ground for the Levee, the trunks to be burned with the rest of the clearing-spoil. During high-water the falling of a tree, from either side, across the embankment, will cut down through the crown at least several feet. I have known one instance where a large cotton-wood (4½ feet in diameter) cut a 5½ feet Levee to its base. If a similar cut should occur during the flood season, on a high Levee, the water admitted through the gap so made would form a crevasse, sweeping away large lengths of the Levee, inundating the adjoining plantations, and for the season of its occurrence, destroying the object of the whole system of Leveeing to the people and property of hundreds of square miles. But, besides the avoidance of this danger, the removal of those leaning trees, at first, is in fact less expensive than when they have fallen.

Road-crossings are very frequently cut across Levees, in Mississippi, and elsewhere, during low-water. The planter immediately concerned is expected to see, at the proper time, that such a cut is duly filled; but in some stretches of Levee, it often occurs that what is every body's business is no body's. Besides, this liberty with the Levee is bad in principle; for it points directly to impunity, for infringements on the sanctity—so to speak—of the work in less dangerous particulars. Rows of Osage-Orange, or other hedge-shrub set along the base of each slope, will save the embankment, better than all the restrictions of law, from injury by either man or beast. In the absence of these hedges, however, it may be suggested that, to guard against the cutting of roads across the Levee, the best course would be, as in Holland, to raise the natural surface of the road-way by embankment from each side in easy slopes, to the top of the Levee. To slope down the Levee-level at a rate of even 20 to 1 to the level of the road on either side crossing it, will, in general, require comparatively little work, the base of such an extra bank, exclusive of one-half the base of the Levee, being for crossing a Levee of 10 feet high, but in all some 370 feet. This road-way is, in Holland, termed "Ramp." The cost of this extra work is but small for securing the advantage of placing the continuity of the Levee beyond the accidents of local carelessness—of placing the important principle of its inviolability beyond the infringement of popular *necessity*.

Levees across creeks or bayous are very often made wide enough on top to constitute a roadway. The inviolability of the Levee comes in here again to object as an important principle against this practice. Besides that the course is objectionable on the ground of economy. The grassing of the crown as a saving of wear and tear is, with the supposition of the roadway, out of the question. If cattle-trespass on the slopes is to be excluded, it can be excluded with a roadway on the top by only the cost of four instead of two rows of hedges, two on each side of the roadway and two along each slope. The wear and tear of the whole Levee with the rut-cuttings of wheels in wet weather, and the slope-breakings of horse-hoofs in wet and dry weather, would make this roadway tell heavily on the Levee-account in increased outlays for maintenance. These objections apply with equal force to employment of Levees for roadways, whether along their whole length or for any part of that length. In the case of bayou crossings, or the crossings of other deep breaks in the general surface of the back-land, the roadway may be combined with the Levee without going to the extent of making them perfectly identical. The following section shows a method of combining the roadway with the Levee at deep-crossings with, in general, less work—a method, too, removed from the objections urged above against using the crown of the Levee for the purpose.

The Levee here is assumed to be protected fully from trespassers by the hedge-rows shown at a and b, the whole surface

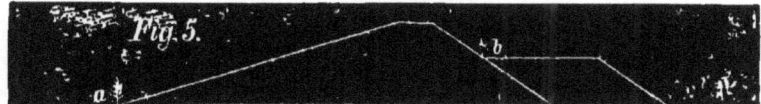

intermediate between these being protected by the proper coating of grass. The roadway, of course, is situated on the dry side, and as an extra to the Levee, may be left to the parties concerned in it as a road for its maintenance. Three-fourths of it may be washed or worn away without any inconvenience to the Levee. Improvements of this sort on the present practice must be regarded with the consideration due to everything pointing to a saving in public outlay—in taxation. The general question of making Levees the site of the roadways required for the traffic and travel taking their direction, is met in the foregoing remarks on Levee roadways. The gravest objection, as has been seen, applies to the location of roads on drainage embankments. Occasional travel even is so injurious that it ought to be avoided by all means—where more efficient means are not employed by the collection of brush, briars, or other obstacles across the whole extent of the bank at intervals. These impediments are absolutely essential in new works on new locations; for few men will be so scrupulously observant of law as to ride through the tangled paths of a swamp when they may choose, in their stead, the open and unbroken smoothness of a Levee. The enforcement of penalties, under such circumstances, is difficult. The best remedy in all such cases, after that of impervious hedge-rows, is—and for travel and traffic only, it is a perfect remedy—the construction of a roadway within the Levee at a distance sufficient to save the berm from the contingency of encroachment by either hoof or tyre. This, with brush walls drawn across the bank at intervals, will save the Levee from all damage except that arising from the trampings of cattle and the "rooting" of hogs. The Osage-Orange, however, is infinitely better than an army of police and a volume of penal laws, for the protection of embankments from all trespass.

Excavations of the ground outside a Levee is objectionable. In sandy or other weak earths it is even worse so than in clays. Under any circumstances such cuts ought to be removed as far as possible from the berm of the Levee; but never less than ten feet. The "pits" dug in such positions ought not to be continuous; but ought to be divided from each other by walls preserving the continuity of the natural surface. Separated thus from each other, those excavations will fill up the sooner under the depositions of floods. These breaks in the external cuts will also prevent them of becoming channels of flow; and thus guard against the creation of avoidable current washes on the slope and berm of the embankment. The slope of those external pits should, on the side next the toe of the Levee, never be vertical, but always dressed off at an angle fully equal to that of the adjoining Levee slope. Left at a less slope, the pit-banks may fall in, and the falling so occurring will then advance until finally it shall have undermined at least a portion of the Levee itself. Trifling as this detail may appear, it is urged here as one which practical experience has pointed out as of great importance. Exterior excavations, then, for the construction of Levees should be made in pits separated from each other at intervals by walls, the inner slopes of these pits being never nearer the Levee base than ten feet, and never of a more rapid angle than that of the water-slope of the adjoining Levee. These precautions ought to be laid down expressly in Levee-contracts. Carried out practically they will save the work from the contingencies of its first and perhaps second year of trial; but after that, the *foreshore*—the ground outside the Levee—will "warp" or "silt up" under the flood-deposits until the resulting elevation going on to the full height of high water mark, the foreshore side of the embankment will cost little or nothing for maintenance.

Large Levees require in their construction especial care. As a general rule it may be observed of those works that the heavier they are the weaker is their natural foundation. The twenty or thirty feet Levee in all cases within my observation implied a Levee, whether across swamp, bayou, or "old bed," having for its base a soil no stronger than shifting sands or watery puddle. All such Levees, on this consideration and on the further consideration of proper compactness and strength, ought to be carried up in *regular layers* of earth, each layer "dished" out from the centre and tramped over by the hauling necessary for the next succeeding layer. These layers should not exceed three feet in thickness. Crusts thus carried up one after another from the base, assist to distribute the pressure of the whole equally over the whole base; and thus in the case of weak foundations assist largely in the stability of the work. Hollowed out from the centre—"dished"—these layers or crusts of tramped earth fitting each into the one below it cannot *shift* under the lateral strain of high-water. After ensuring safety at first by carrying the bank up thus in layers, the whole becomes subsequently one solid and settled mass. The settlement allowance in banks so constructed is merely nominal. In connection with the stability of Levees across *out*-discharging bayous, it may be observed here, that that stability is sometimes threatened, after all proper precaution in construction, by the accumulation of water on their inner side. The bayou having discharged outwards a mere *trickle* perhaps at the time of stopping it, soon accumulates to a considerable body of water, until finally the whole becomes ponded up against the sides and bottom of the bayou and the inside slope of the Levee to the height of the bayou bank. The additional load on the foundation over and above the Levee and the river flood-water, is in itself objectionable, though perhaps not quite so much so as might appear at first, when it is recollected that the distribution of the weight, front and rear of the Levee, would save the danger of "bulgings" up at the toe of the Levee—the place of these bulgings being always occupied by corresponding sinkings of the embankment. But the greatest objection of such pondings up is their continuance; seeing that the longer they continue the more thorough and the deeper is the saturation of the underlying earths; and the more thorough and deeper their saturation the greater the extent and the degree of the weakness or "meltings" of the earths in the foundation of the Levee. In order to guard against this evil, it is necessary that small cuts be run up through, or out of, such bayous to such a point as may be necessary to divert their drainage into the general outfall of the surrounding country. All Levees across *out*-flow creeks or bayous have been observed, when this precaution has been neglected, to sink into their foundations; and as a consequence to cost more than otherwise for their maintenance.

Crusts of earth have been referred to already as means for the distribution over a wide surface, of superincumbent loads. In the case of a "muck ditch" cut along the site of the Great Yazoo Pass-Levee of 1855—cut against the express direction, if memory serve me truly, of the Commissioner and the Engineer—the advantage of an unbroken crust of this sort was illustrated very strikingly. The division by the "muck ditch" having taken place under the line of greatest load, that load, pressing on the lips of the cut vertically, acted on them as remarked in such cases on a previous page, with a leverage, until driven down step by step the crust must have become, by the consequent bending, broken near the toe of the Levee on each side. Two distinct pieces of crust were thus, in all probability, driven down angularly into the thin matter underneath; and thus, instead of using every precaution to preserve the natural crust as a grillage under the bank, the Levee was left to seek its foundation as best it might amongst quicksands and fluid puddle. The weakness of the foundation told itself accordingly by not only the sinking of the crown of the embankment, but also by the bulgings of the slopes and the spreading of the base—"a corduroy" or causeway of logs near the inner *talus* having been forced by those spreadings from a straight line into a succession of zig-zags. Alarmed by the sinkage of his work—a sinkage that, according to his estimate, was made naturally enough by one in such a position, to cover all the losses consequent on his own mismanagement—the contractor became unmanagable. The Engineer recommended him to open a ditch of a few feet wide and of five or six feet deep inside and parallel with the Levee, in order, by filling up the same with the trunks of the young cotton-woods cleared from the base of the Levee, to present a breast work of equally bearing resistance on the dry side to the spreadings of the base—the then-standing flood-waters on the wet side offering sufficient resistance to spreadings or bulgings on the wet side. A deep pit of sand furnished convenient drainage for the ditch so suggested ; and the excavations of the ditch would have been available for the completion of the embankment according to the contract. Instead of adopting this course, however, the contractor attempted to carry up the sinking Levee to the contract-height by removing the masses of earth bulged out on its sides and spread out beyond its base to the crown, heedless of the remonstrance of the Engineer that in doing so he was merely revolving an endless chain. Every yard taken from the bulgings on the side and removed to the sinkings on the top, was speedily replaced by another yard on the sides, the top remaining in *statu quo*. Finally, however, the work was resumed at the base and carried up in closer conformity with the slopes ; and, all parts of the cross-section thus newly loaded, the crown carried up finally in "a comb" to exclude the flood then standing some 20 feet on the water-slope, the whole presented sufficient solidity, under all the unfavorable circumstances that attended its construction, to have dammed back that year's flood. This case fell within the practice of Mr. M. B. Hewson. Another case of weak foundation for a Levee came under the observation of that gentleman in the construction of the great Levee across the mouth of the Old Lake of Oldtown, in the State of Arkansas. The site of this Levee had, within the memory of men living in the neighborhood, been the bed of the Mississippi River; and as such may be well supposed to have presented a foundation of a description the very weakest. A shallow stream running across the proposed line of work, the undertaking had not the advantage in some places of even the hard crust of the Yazoo Pass. The irregular course pursued in carrying out those works on the Mississippi confines generally its restriction on the contractor to the height, width at base, and width of crown, the means by which he fulfills those conditions being questions for only his consideration. Acting on the part of the State, Mr. Hewson had no voice as to foundation or any other obvious preliminary in a proper construction of the work. The contractor accordingly dumped in his earth without any preparation of the foundation; and counting on pay for every yard so dumped, carried up the work by, simply, force of purpose and labor. The centre *frequently* sank into the foundation. Standing sometimes for a couple of hours at its full height, it would drop down suddenly from 5 to 10 feet. The base spread to an incredible extent at all points but one—that one being loaded heavily with an interlocked and tangled mass of logs, branches, and trunks, removed in clearing the site of the work. The spreadings having reached their utmost at all other points, the work was being carried up along the *loaded* length of the slope, when suddenly the tangled mass of timber loading it was torn asunder with *a loud noise*, and *shot* for some distance from its original position. The bulging that took the place of this weight on the base of the Levee was found by measurement to have been some 4000 yards! Such are the forces exerted by such Levees; and such the character of their foundations. The "sinkage," as it is locally termed, was, however, in the case of the Yazoo Pass-Levee, a mere trifle in comparison with that at Oldtown.

Weak foundations occur in the case of Levees, in only the cases of those important works that may be considered the Keys of the whole. The Yazoo Pass embankment once swept away, the whole extent of Levee remaining in the County of Coahoma would become, virtually, valueless to the parties living behind it as a protection from inundation. The undertaking of the Levee-system at all, involves, therefore, the necessity that all those more important points of the system be executed in a manner to insure, at least, as high a degree of stability as any of the less important parts of the system. With some 30 feet of water standing on its outer slope at flood-time, its inner slope resting on the bed of a channel, of like depth, constituting the arterial drain of the back country, the failure of one of those Key-works of the system, when occurring to even the smallest extent, involves its total destruction. The rush of water through the whole width of this Key-levee, under a head of some 30 feet, sweeps into the back country in a foaming torrent ; the whole system of back drains becoming, in the first place, suddenly charged to the lips ; and then, all the overflow passing off upon a surface deprived of out-fall, the country behind the Levee becomes, to a greater or less extent according to duration of the flood-level in the river, completely deluged. On the other hand, a "crevasse," or breach in the general line of the Levee, may not only be stopped altogether before it arrives at any considerable width ; but at the worst, the depth of its out-flow not exceeding a few feet, the back lands pass off the water, through their system of back-drainage, at a rate, if not even quite as rapid as the in-flow, quite rapidly enough in the generality of cases to prevent the engorgement of the back-drains before the fall of the river-level. A breach in the Key-levees, then, involves a certainty of wide inundation ; while a breach in a less important part of the system leads to an inundation under the worst circumstances, limited in its extent, and in its duration brief. But this is not the only reason why it becomes necessary to construct Key-levees with special care. Costing, according to their present mode of construction at a rate so high, in some cases, as upwards of $60,000 a mile, whereas, the generality of Levees do not cost over $2,500 a mile; a breach in one of those heavy works—leading as it always does to its total demolition, results in a very serious loss of money. The destruction of Yazoo Pass-Levee was as great an injury to the treasury of Coahoma County as would have been the destruction of the whole embankment, from the junction with the Pass-levee to a point as far South as Friar's Point! The Commissioner was very much censured by parties interested in the stability of this Pass-levee, for his special outlays on construction of this work, and for his rigid enforcement of the conditions set forth in the contract for securing that stability; but how thoroughly rebuked his short-sighted, and, perhaps, factious censurers must have felt, when they discovered to their cost, by the destruction, in 1855, of that most important work, that in all his care and all his "harshness," Col. Alcorn was pursuing the course as a public servant, of courageous honesty and enlightened carefulness. The truth is, Col. Alcorn felt at the time, that his duty in the case of the Pass-levee was rather under-done than overdone; and in conversation with Mr. M. B. Hewson, on the subject, frequently referred to the embarrassment in which he was placed in the case, by the absence of a sound and intelligent public opinion as to the conduct of those works. The purse-strings being in the hands of the tax-payers in the case of Levees, it is of importance to their best interests to place such measures as are necessary for the proper construction of those works, under the endorsement of their understandings. Having with this view referred to the special importance of such works as the Yazoo Pass-Levee, and the Old Town-Levee, &c., it is proposed, now, to offer some general remarks on Levee-foundations.

The more important portions of the system of river-embankment in the case of the Mississippi, rests, as has been said, on foundations of puddle or quicksand. Continuing to dump in earth into embankments on such a foundation, is found in practice to result, after a greater or less waste of earth, in compressing the foundation downwards and outwards to a compactness sufficient for the resistance necessary to sustain the intended Levee. The Levee accordingly stands—its height, crown, and side-slopes being in perfect accordance with the contract. This is all the public expect; and unfortunately, with their present views, this is all they will sustain their Commissioner in enforcing. The flood, however, rises on the face of this Levee—10 feet, 15 feet, 20 feet—even 25 feet, and a load of 600 lbs., 900 lbs., 1500 lbs. becomes thus added to every square foot of the outer half of the Levee base; while no corresponding weight on the inside is available for establishing a counterpoise in the watery material of the foundation. The compression that the foundation underwent, originally, before the Levee attained the required height, was the result of the weight of the earth employed in its construction; and it is only reasonable to infer that, with the same foundation, a further load, whether of earth or of water, will occasion a further compression. Additional sinking, or "canting" under the outside slope, and additional bulging, or spreading of the inside slope, is a natural result under such circumstances; and the special weakening of foundation under the special saturation of a superincumbent head of water, combining with the other natural result in the case, no surprise ought to be felt that the great Levees, constructed after the general practice on the Mississippi, should be swept away before high floods. To construct those works properly, then, requires special steps in reference to the strength of their foundations. Brush makes a very good foundation in weak soils. McAdamized Roads have been carried through otherwise impassable marshes in England, on foundations of brush laid in considerable thickness upon the surface of the marsh. It is extensively used in Holland, and the "Low Countries," to strengthen the foundation of heavier embankments than are likely ever to occur in practice upon the Mississippi bottom. In Ireland, heavy embankments of the Grand Canal, and also, embankments of the Great Western Railroad have been carried in several instances across deep "flow-bogs" on brush. The Grand Trunk Railroad, of Canada, has a bank some thirty or forty feet high, across a deep and wide marsh, sustained by a brush foundation. Several such instances of the use of brush might be mentioned here, to show how useful for the purpose of Levee-foundations is a material that, along the banks of the Mississippi, may be obtained without stint or trouble. The branches of the trees cut off for the purpose, should be laid evenly across the base of the Levee, in layers 24 inches thick, the direction of those in each layer "angling" across the line of the base, those of the layer next above being laid "angling" in the other direction. Two layers are quite sufficient for ordinary heights of bank, and ordinary weakness of foundation; but in other cases, it were better to lay three layers or even four. "Old beds," such as the Yazoo Pass, or Old Town Bayou, should in all cases be brushed with four layers, compressible to a thickness of, at least, six feet, each layer having its branches laid across the line of Levee, askew—the second layer crossing that of the first, and so, also, with the others. The brush should be cut off regularly, so that it would never extend on either side within ten or twelve feet of the toe of the bank; pressed for even its own thickness into the foundation, and the brush covered up completely, the embankment resting upon it will not, necessarily, be open to the objection of leakage. Brushing, properly and carefully employed, in even the highly unfavorable circumstances of the Key-works of the Levee-drainage, will constitue a perfectly stable foundation. Fascines are sometimes employed in the foundations of embankments; but, while much more troublesome, are not so efficient as simple brushing executed properly. Fascines may be described as small bundles of brush, each tied firmly like a birch-broom. These are laid down one row across the other for foundations of banks; but all the advantages of their compactness may be secured in brushing, by selecting the brush carefully in the first instance—long, straight, tough, and sufficiently light—and in the next instance, pinning it down occasionally by wooden forks to the *ground* in the first case, and to the layer below it, in the next. More perfect continuity latterally and longitudinally may be obtained with the simple brush than with the fascines. Sand is another material most available along the Mississippi, for the purposes of artificial foundations. Loose in its parts as it is, sand is not supposed, generally, to be capable of constructing a mass of such stiffness as to distribute over its length and breadth the pressure of a heavy load. This, however, is the fact. Like water in other particulars, it is especially like water in discharging its pressures, under certain circumstances, latterally as well as vertically. This property of sand has lead to its employment in foundations as a substitute, in certain cases, for piles of wood and of iron. Wooden piles obtain their bearing mainly from the resistance presented to their section; but sand-piles, in addition to this resistance, are found also to present a further resistance along their sides. Friction or adhesion, as it may be, this increased sustaining power of the sand-pile has been found highly useful in the preparation of foundations for heavy Engineering works in soft and deep alluvium. Where the base is not so weak as to require the use of piles, artificial foundations of ample strength are sometimes obtained by spreading over the natural surface a thick and uniform coating of sand. In wet situations, however, hydraulic lime is sometimes added under this practice to the extent of one-seventh the bulk of the sand; and as such an addition would, in all likelihood, be found necessary in the case of the great embankments of the Levee-system, the use of sand for artificial foundations for Levees, must, on the score of economy, be confined to piling. Brush, however, is in all cases the best material available along the Mississippi, for the preparation of a Levee-base; but before loading an unusually weak base with an unusually heavy embankment, it would be well, in addition to the brushing, to sink one row of sand-piles immediately under the intended site of the crown, and two other rows on each side of it, the piles in each row alternating regularly with those next it, thus:

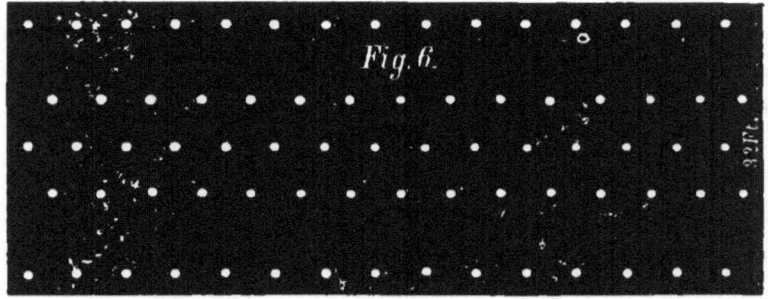

These piles ought to be about from 1 to 2 feet in diameter, the centre row being placed at intervals from centre to centre of say six feet, the next row on each side being parallel with this centre row at a distance from it—from centre to centre—of about 8 feet, the intervals between the piles of those second rows being from centre to centre 8 feet. The outside piles ought to be sunk at a distance from each other and from the adjoining rows of 10 feet from centre to centre. These piles may be put in by several methods. A light lift-ram being employed to drive into the intended place (where the same may not be done by a heavy sledge) a wooden pile of the intended size, this pile, after being thus sunk to the required depth, must be withdrawn, and the hole filled in rapidly with sand for the purpose, this sand-filling being compressed at intervals as it progresses, by blows of a rammer. By the time this sand pile is thus filled in and compressed, the driving-party will have been placed ready for work at an adjoining pile; and thus the whole area will be piled, each pile being completed before the driving shall have been commenced for that adjoining it. Six or seven feet is quite sufficient as a depth for those sand-columns. The best method, however, for sinking those piles where the extent of the work would justify the outlay, would be by an iron cylinder, furnished on the inside with screw threads one-half the diameter in width; for this might be sunk and raised without machinery; and permitting conveniently of increased size, is well adapted to piling with sand columns of two feet in diameter. The larger and the closer the sand piles the more thoroughly do they pack the material of the base, and the more effectively do they increase their own bearing-strength. The regular mode for sand piling is that pointed out here; but the rough mode of doing every thing in Leveeing will probably substitute, in sinking those piles, a pole forced into the ground by manual strength, and, in withdrawing it afterwards, worked around its point until the hole becomes sufficiently widened. The bottom of the pile, it must be recollected, will be very small under this proceeding; but by ramming the sand thoroughly into the hole—water or no water—sand-piles will, under even such poor construction, assist largely in compressing the soft earth around them, and in supporting the load of a heavy embankment. Be the mode of construction for those piles then what it may, they are recommended in all cases of unusually heavy embankments and unusually weak earth; and when even partially employed are highly valuable accessions to "brushing" in artificial foundations. The popular understanding must, however, be satisfied in reference to every reform in the construction or management of Levees; and as the association of strength with a "foundation of sand" conflicts with all the previously formed views of that popular understanding, it is well to sustain the use of sand-piles—as has been done in the case of the recommendation of brushing—by reference to specific practical tests. In reference to the general fact of sand being a strengthener of weak foundations, it may be observed that it has been employed successfully as such, under heavy masses of masonry at Geneva in Switzerland, at Bayonne and Paris in France, in India, in Surinam, and doubtless in many other places where its use has not fallen within the knowledge of the writer of this. In the "Annales des Ponts et Chausees" —the Reports of the Board of Public Works of France—for 1835, a complete account of the use of sand in foundations is published; and in order to satisfy the sceptical planter as to the utility of that material in foundations, the following extract is made from that report in the translation of a professional paper on the subject, by a member of the British Corps of Royal Engineers. "On a very soft soil nine piles about, 4 feet three inches long and 8 inches in diameter, and distant from centre to centre about 16 inches, were driven with a monkey weighing about 2 cwt. falling from a height of 3 feet 6 inches; the driving was continued till the piles only yielded about ¼ of an inch at a stroke; upon these piles a load of 20,000 lbs. was placed and the settlement amounted to about one-fifth of an inch. These nine piles were then drawn; and the holes in the soil filled in with sand; 16 more piles were driven in the same way so as to occupy a space of 6 feet square; the ground was then well rammed; and a mass of masonry similar to that in the former experiment was built and loaded with lead as before:

Under a weight of 1050 lbs. the settlement was 1-25 inch.
    "   " 2100 lbs. "   " 2-25 inch.
    "   " 3150 lbs. "   " 3-25 inch.
Which increased to   4-25 inch.
Under a load of 18 tons the settlement was   1-5 inch.
21 tons made no sensible change, and 30 tons increased the settlement about 1-50, and after a month the total amounted to   3-5 inch.

A well about 12 feet deep filled up with silt and clay; after having removed about 16 inches of soil from the surface the under stratum was found quite soft, a Ram penetrating 6 inches at a stroke. To harden this soil, 25 piles were driven about 4 feet 6 inches long each; this forced the soil up about 16 inches above the previous level; the driving was continued till 20 blows of a Ram weighing 2 cwt. let fall from a height of 3 feet only, caused a pile to penetrate about 4 inches, which took about 40 minutes' work. After having driven all the 25 piles and levelled their heads, they were loaded as follows:

    12 tons caused a settlement of about 1-20 inch.
    18 tons   "   "   1-10 inch.
    And in three days this increased to 1-5 inch.

These piles were then drawn, and the holes filled with sand, which was well rammed, and which ramming caused a barrow full of earth to bulge up between the holes. On the ground thus prepared a mass of masonry was constructed and loaded with lead as before, and the settlement was as follows:

    15 tons caused a settlement of 1-10 inch.
    29 tons   "   "   2-5 inch.

These weights were placed in April and remained on till December, when the increased settlement amounted to $\frac{3}{4}$ of an inch. The load being reduced to 10 tons, no further settlement took place between December and May." Other cases of the employment of sand-piles are given in the same reports; but this is sufficiently specific and forcible to satisfy any reasonable doubt as to that employment in the foundation of Levees. Some of the cases given have been in situations where the silt and alluvium was over sixty feet in depth. Let the holes then be opened as they may, pack them well with sand as close to each other as the circumstances of the case may demand; and the result will be, in all cases, an accession to the strength of the foundation, and, therefore, to the stability of the Levee.

Brush and sand combined will undoubtedly sustain the heaviest Levee under even the most unfavorable circumstances ; and the most available and cheapest materials for the purpose being thus sufficient for the requirements of any case, it is unnecessary to consider any other methods than those based on the use of these for accomplishing that important object in Leveeing, substantial foundations.

The difficulty, however, of meeting every particular of improved construction in the Levee, lies in the sneer of the purse-holders in the case at what it calls "College" Leveeing. But the taxpayer who is enriched by the Levees of the Mississippi will make a grave mistake in assuming himself exempt from the hard lessons that have taught people placed under like circumstances in other countries, the wisdom of their present practice in water-embankments. In order to urge on the popular judgment of the valley the importance of the suggestions made above for fencing, grassing, and otherwise protecting the work, after its construction, for mucking, layering, brushing, and even piling, in the course of its construction, it may be well to add here a few instances of the costliness and care involved in the case of European and Asiatic water-embankments. Touching the dimensions recommended above, it may be observed that Sir Cornelius Vermuyden gave the embankments of the Welland in England for a height of 8 feet, a base of 70 feet; and that of the Ouse—with a crown of ten feet—for a height of 8 feet, a base of 60 feet. The Ouse at Weisbach has for its embankment a height of 10 feet carried up on a base of 100 feet. Sir John McNeill, in the drainage of Lough Swilly in Ireland, gave his embankments a base of from $5\frac{1}{2}$ to 6 feet for every foot of their height. On the Lower Damoodah in India, the "Bunds," as "Levees" are there termed, have a base for 4 feet high of 23 feet. In reference to the inequality recommended in the case of the dry and the wet slopes of Levees, it may be added that the practice of this inequality is universal. Sir John McNeill—a name distinguished highly in his profession—has given the Lough Swilly embankments, already referred to, an outside slope of from 3 to 1 to 4 to 1, while the inside slope is but 2 to 1.

Sir John Rennie reports the construction of an embankment in connection with the Commissioners of the Nene Outfall, the dimensions of which were 5½ to 1 to seaward, 3 to 1 to landward, the top being 4 feet broad. Arthur Young, in his agricultural reports, mentions several cases of embanking, one of which he says was erected in 1800: the dimensions were for the sea slope 4 to 1, land slope 2 to 1, 12 feet in height and 4 feet broad on top. The Dikes of Holland—constructed principally as defences from the sea—are generally raised 30 feet above the ordinary level of the country. Banks have been occasionally made twelve feet wide on top, and carried 2 feet above the high-water mark, they were in some positions turfed and strengthened in various places with stakes, or piles and planking. Sometimes banks were formed by driving rows of piles or stout stakes parallel to the river at distances of from 2 to 3 feet apart, and after uniting their heads by a plank, or weaving rods around them, the parallel spaces between were filled up with chalk or some other hard substance. Banks made of sand, in which twigs of brush-wood are placed horizontally and clayed properly with from 1 to 2 feet thick of clay, are found to stand remarkably well. Coroboratory of the stress laid upon the selection of clay for Levees, it may be observed that in Holland, where the system of water-embankment is carried out under all the experience of centuries of disaster and destruction of life and property, and where the adoption of the best mode in each particular is a question of national concern, sand in the site of the Levee is rejected when better material *cannot be obtained short of a haul of even five miles.* In Europe, no trouble or cost is spared in the construction of water-banks to make them perfectly water-tight. Carrying up within the work a wall of puddle, a practice which is known by every Engineer to be universal in the construction of all water-tight embankments, does not require for its endorsement an isolated instance like even that in the banks of the Swilly-drainage in the practice of Sir John McNeill. The care recommended in reference to the foundations of Levees is borne out by general usage in such works—in England, in Ireland, in France, in Holland, in Germany, in India. The preservative measures urged above are but modifications based on universal practice. In England and in France water-banks are regularly "turfed;" and in weak places protected with even stakes and piles. In Holland the water-banks are protected on the outside by a strong coating or matting of flaggers and reeds; and on the inside are sustained by piles and planking, the slopes being coated thickly with grass. In the Swilly works, as illustrating the latest practice and highest experience in Europe, it may be remarked that the land-slopes of the embankments were all covered with turf; the water-slopes having been protected with a facing of fascines six feet thick at bottom and 4 feet thick at top, these fascines being laid in an oblique direction in the slope and fastened thereto firmly by forks of iron. Reason then has been first appealed to here in urging the adoption of the care recommended in the construction and protection of Levees; and these last references to the subject, show that the suggestions of reason in the premises, are fully endorsed by the general *practice*. But another point for viewing the subject may be illustrated by a reference to the cost of Levee-*maintenance*. In Zealand the *maintenance* of their embankments—300 miles in length—cost them *annually* $800,000! The *maintenance* of the embankments and the regulation of the water-levels in Holland, cost the enormous sum of $3,000,000 *a year!* Sea-banks those works in both Zealand and Holland chiefly are; but, on the other hand, it must be observed that in reference to the cost of their maintenance they have been constructed with the best material in the best manner, and their preservation has been guarded since the time of their construction, with all available preservatives against decay. River-embankments, it is true, the Levees of the Mississippi are; but it must be recollected that, until the management of one portion of those works by Col. Alcorn, very little care whatever had been taken in their construction, and equally little in their preservation. The *maintenance* tax then must continue a heavy burden for some time to come on property-holders within the Levee; and this tax is subject to reduction in only the amount of care expended in constructing new Levees and in preserving both new and old. The method of construction and the means of protection after construction recommended above, are thus seen to be means adapted to *reduction of taxation* for Levee purposes on the parties chargeable with their maintenance. In this point of view then it is hoped the popular judgment will hesitate before it undertakes to sneer away recommendations so influential for public good as mere " College " nonsense. These remarks are intended not for the *intelligence* of the valley; but for the guidance of that popular mind which may stand at the ballot-box, an impassable obstacle to even such a bold and talented reformer of the Levee-system as the distinguished gentleman to whom this volume is inscribed.

CHAPTER V.

HIGH WATER MARK.

The municipal line is supposed, under the present law in Mississippi, to make each down-stream County higher than the flood that may inundate the County above it. Boliver is, therefore, assumed to have no concern in the drainage of Coahoma. Engineering, then, must trim its practice to meet the absurd system of mere County jurisdictions. The wide revision suggested hereafter, for the correction of flood-levels in Levee-surveys, cannot be carried out satisfactorily at present; and hence does it become necessary to offer a few remarks on flood-levels in connection with the cramped working of the existing Levee-law. Simple as it may appear, the establishment of the High Water Mark along the lines of Levee, is very often a source of difficulty even to the skillful, and of error to the unskillful administrator of Levee *Engineering.* * And

---

* Illustrative of the errors in construction resulting from want of skill in determining High Water Mark, the following extract is made from the Report of 1856, by the then Commissioner of Levees, in Tunica—Judge Hardeman:—" The profile of Mr. Hewson's survey, &c. * * * * On being with him during the survey, and his taking the field notes of the same, it has been clearly indicated to me, that *all the Levees heretofore built in the County*, —except that portion built and repaired the past season—are from *one to two feet too low*, &c. These Levees should claim our first attention and, if possible, be repaired the coming season." Judge Hardeman might have even stated that the profile from which he speaks, showed some stretches of the embankments so much *above* High Water Mark as 5½ feet; while it

error, be it recollected, in this particular is the most dangerous that can happen in all the facts affecting the system. In the first place, it requires great caution in accepting testimony, generally more or less hazardous and loose, as conclusive. Checking the flood-levels above and below the point of difficulty in such a case, is the only means of passing from such conflicts of authority to fact. Nor in making such a comparison must it be concluded that the rate of descent of flood-water is always that of uniformity. Across a bend it may be very rapid; whereas, the line of flow crossed by a dense forest, thicket, or cane-brake, the flood-line, for a greater or less distance up-stream, will be either that of exceptionally slow descent, or of even a dead-level. But further difficulties apply to the acceptance of local evidence as to High Water Mark. In Levees of wide fore-shore, it is a great, sometimes a dangerous mistake to accept as absolute the High Water Mark cut on a tree on the spot, during the flood of 1844, or 1850, by even that oracle of flood-facts, an "Old Surveyor." The "Old Surveyor," doubtless, may have even made the mark at the exact height of the flood-water, then and there; but this amounts to simply nothing, when it is recollected that the flood-water at the point in question was the flood-water of an *outflow*, and that the river-bottom, while having a fall of 7 or 8 inches per mile along its axis, has, *across that axis*, a fall of 5 or 6 inches per mile. The Old Surveyor, in short, forgot that, there being a fall in the outflow and also in the ground from the River-bank to his flood-mark, the damming back of that outflow would throw up the flood-level proportionally higher. This consider-

showed other lengths, so much as eighteen inches *below* High Water Mark. The report made to him on the subject, holds in fact these words:—"The old Levee, for 27¼ miles, requires an average additional height of 1·55 feet, to bring it up to the required height of 3 feet above High Water; in some cases, this length of Levee was found *to dip below the High Water-level*—each such dip involving, in flood-time, certain destruction to considerable lengths of the work above and below it."

ation is highly necessary in ascertaining the *true* working High Water Mark. The ready-made Engineering of the first Levees—if indeed, it be any thing more intelligent in some places to-day—overlooked the fact that the flood-marks across the bottom follow the combined slope of ground and outflow; and in consequence of this extraordinary error, many a mile of Levee has been swept away, many a dollar, in both scrip and cash, wasted. In practical illustration of the difficulties of fixing the High Water Mark—a duty for which *every* man along the Mississippi regards himself perfectly competent—it may be observed here, that the High Water Mark taken, on all the local tests, at a point on the upper reach of "Old River," at Port Royal, in Coahoma County, Mississippi, was higher than the High Water Mark taken with like care, at a point half a mile down-stream, by so grave a difference as $4\frac{1}{4}$ feet! Indeed, the evidence available in the case is so loose and uncertain a guide; and an error in that guidance, involving the destruction of the Levee, it is highly important, in order to proceed under all the available lights with safety and confidence, that the selected levels of High Water Mark be compared one with another, along each whole drainage district; and, finally, be revised by comparison with the selected levels corresponding to them on the other side of the river. But this necessity supposes an improved system of Levee-law and Levee-administration. For the future, however, it is highly important that the public attention be directed to the wisdom of recording, as often as possible, along the river, the height of each year's flood. While the memories of parties living along the bank, on both sides, are fresh with marks of the late disastrous flood, a movement just now would be well timed for the commencement of such a system of record from end to end of the inundated shores. Well-ascertained evidence of this sort may be fixed at once, by the leveller; and after comparison and selection of all the facts, he may transfer the revised flood-heights to a series of

Bench Marks, sunk for the purpose, at intervals, inside the Levee. These Bench Marks should be driven down, firmly, three or four feet into the ground, so as to guard against their being broken or sunk; and when their levels may have been duly ascertained, that of each in reference to its flood-level should be marked in red chalk or paint on one side, the number of the Bench Mark being marked, likewise, (in order to identify it) on the other side. This use of Bench Mark-stakes is universal in the Engineering practice on the "Dikes" of Holland; and like most of the usages established for the conduct and maintenance of those works, is highly applicable in the case of the Mississippi embankments.

CHAPTER VI.

LOCATION.

The location of a line of Levee is a consideration involving permanence—involving economy of construction and economy of maintenance. Large sums of money have been expended in Levees which, in several instances, have *within twelve months of their construction fallen into the river*. The cause of this has been ignorance or carelessness in determining the location. Private interest, however, is very often a disturbing influence in forcing the location of Levees from the line of safety and economy. A Planter has frequently been known to be so short-sighted as to have urged, and in fact obtained, the location of a Levee around three sides of even a "turnip patch" rather than consent to the necessity—to himself as well as to the general public—of locating that Levee in continuance of its proper alignment directly across that "turnip patch."* The increased cost

---

* In the address explanatory of the causes of Levee failure during the late flood of the Mississippi, the Coahoma Commissioner, while putting the scientific considerations of the case in good popular terms, caps those considerations by reference to late practical experience; on pages 17 and 18 he says: "Motion, whether of solids or fluids, naturally follows straight lines; and all deviations from that law are accomplished by an expenditure of impulse on the object occasioning that deviation. A sudden turn in a stream concentrates the whole energy of the fluid-motion on the one point, occasioning that sudden turn, hence the danger of all sudden turns in the Levee. In the original locations of the Levee all these laws of motion were violated; *no regard whatever was paid to the alignment;* it was made to *wind itself around*

of constructing the embankment to meet this gentleman's narrow-minded views, as compared with the cost of constructing the embankment on its *proper* alignment, has very often been ten times greater than the whole value of the additional piece of ground he had, by forcing the Levee out of its course, succeeded in enclosing. This, however, is not the only injustice done under such circumstances to the body of the tax-payers ; for in making the Levee on the zig-zag necessary for the gentleman's purposes, that course is subject to the additional injustice of either reconstructing the work on the proper ground when the zig-zag may have fallen into the river, or of flooding the whole back-country when the shock of the high-water current striking against it directly, bursts its way in a " crevasse" through that zig-zag's up-stream juttings. The location of Levees, then, it may be seen from these remarks, should not be a mere matter of random ; but should be determined thoughtfully with a view, in the first place, to the progress of the river whether in " caving" or " making," and with a view in the next place to the obviation of current-shocks.

In locating a Levee, the first duty is the mapping out carefully of the bank ; and, as far as may be done, by a careful sketching of the current-set, the " caving," and the " making." In the case of cavings and makings, every information as to their *commencement*, their rate of progress *inwards*, and their advance *down*-stream, should be obtained carefully from local information and recorded at the proper points upon the map. The cavings and the makings of the bank pass down-stream in a series of waves, period after period ; and, therefore, by ascertaining the rate of descent, the rate of penetration of a

every cow-pen or horse-lot, presenting obtuse angles in the work at many critical points ; and that, too, without any increased strength of embankment at those points. *At many such places the Levee during the late rise gave way*, for the reason, as assigned, it was without strength to resist the current-shock." The most zealous and best informed friend of the Levee-system thus urges and endorses the importance of proper attention to the question of alignment.

"cave," or extension of a "make" at the point of its operation, the location of the Levee opposite that point may be made with a full knowledge of the conditions of its permanence. Levees built one year under such, evidently, necessary precautions, will not be swept into the river within either a few years or a few months after their construction. In order to illustrate this important point more fully, the method of making, and indeed of applying, the notes of "caving" and of "making" as recommended here, may be impressed upon the understanding of young Engineers more readily by a speciality. With this view then is given the following instances. The Chief Engineer of the Mississippi, Ouachita and Red River Railroad, having located the Eastern Terminus of that road at a point which failed to satisfy some of the stockholders, Mr. M. Butt Hewson, then directing the affairs of the Arkansas Midland Railroad, was engaged to report upon the question. The general grounds on which the original location had been based having been taken up by that gentleman as the heads of his inquiry, one of those so made the subject of his investigation was the question of an anticipated change of course in the river by a "Cut-off," opposite Gaines' Landing. Mr. Hewson's report under this head presents the following illustrative remarks applicable to the considerations referred to here as guiding Levee-locations.

"A long professional experience in the improvement of rivers, a somewhat intimate acquaintance with the laws of fluid-motion, and a few years observation as a resident on its banks, of the habits of the Mississippi, fail to place my answer to your fourth question within the limits of exact induction. It is much safer to speculate than to *demonstrate* on the subject of changes of the Mississippi River. I shall, however, furnish you with the facts bearing on your question; and thereby enable you to judge for yourself as to the logical justice of my inferences.

"One general law of the Mississippi River—subject like all general laws to special exceptions—is very plain, viz: the progress of its cavings, like that of its currents, is down-stream. In that portion of the river under consideration, the set of the current from the Arkansas side struck the Eastern bank, some time ago, opposite the residence of Col. Martin; whereas, now, the most Northerly thread of that current does not strike the same bank for several hundred yards *lower* down. So much for the *general* fact of the progression of the *cause* of active caving. I will now call your attention to the present stage of this progression in the reach of river under consideration. Eleven hundred yards below Col. Martin's house, the present caving commences; the Southern limits of this caving is not reached for a further distance of eight thousand one hundred yards still lower. The centre of this existing impact on the bank may, therefore, be deduced as midway between those limits of present caving—that is to say, 4000 yards below the Northern limit of that caving. The force of a current, always a minimum at its outer limits, reaches its maximum in the middle of those limits. Now, the 'Cut-off' suggested, abuts on the bank at 3700 yards below the Northern edge of present caving; and, therefore, the centre of impact, the point of greatest effect, having already, in its steady progress down-stream, passed below the site of the assumed 'Cut-off' for a distance of 300 yards, we may reasonably conclude that, so far as the supposition of this 'Cut-off' is concerned, the period of maximum expectation—of greatest likelihood—is irrevocably passed. The beam that sustains the pressure of ten tons must be supposed perfectly safe from fracture under a like pressure of nine tons. In consideration of these general facts of the case, the inference is clearly opposed to the supposition of this 'Cut-off.' In order to examine the same question from another point of view, I will present an analysis of the evidence as to amount and rate of caving, furnished by gentlemen living on the ground, at the several points along the line of this progressing impact. Dr. Offutt states, that opposite his house (a point above Mr. Daniel's house) the bank has caved 400 yards in 20 years; but at a less rapid rate for the last ten of these, than for the previous ten; and for the last four of these latter ten, still more slowly. Mr. Wallace affirms that the bank at the same point, has caved 100 yards for the last 7 years; and as compared with the gross average of these seven, 'very little' for the last 2 years. At this place the bank has caved:

>Within the last 20 years, at the rate per year of 20 yards:
>Within the last 10 years, at a rate per year of *less* than 20 yards:
>Within the last 7 years, at the rate per year of 14¼ yards:
>Within the last 2 years, at a rate per year of very little.

"Here, then, is a constant diminution of the effect—a diminution in direct accordance with the passing away of the operating cause. Opposite Mr. Daniel's, (a point *above* the suggested 'Cut-off') the bank has, on the authority of Dr. Offutt, caved, in twenty years, five hundred yards; the greater part within the last ten years, while the caving for the last year has been at a lower rate. Mr. Wallace's testimony as to this point, places the cavings at two hundred yards within the last seven years; but for the last two years, very little. These evidences stand thus:

>Within the last 20 years, at the rate per year of 25 yards:
>Within the last 10 years, over 25 yards:
>Within the last 7 years, 28¼ yards:
>Within the last 2 years, much less.

"In this *increase* of effect, up to a certain time, and *diminution* of effect since that time, we obtain further evidence of the Southern movement of the centre of impact. Twenty years ago, it had not reached so low down as Daniel's; and, consequently, did not then produce, at that point, its highest effect;

but as it advanced, its progress is traced in the higher effect of the last ten years; in the still higher effect of the last seven years; and, as it passed further South, its continued progress in the *diminished* effect of the last two years. The point upon which the suggested 'Cut-off' abuts upon the bank has, according to Mr. Wallace, caved one hundred yards within the last seven years; but for the last three years of these seven, at a lower rate: whereas, on the authority of the same gentleman, the bank, opposite Mr. Wilkerson's, (a point *below* the suggested 'Cut-off') having caved three hundred yards within the last 20 years, has maintained a higher rate of caving for the last ten. This point opposite Wilkerson's, coincides with the present centre of impact, as inducted above, from the existing *limits of effect* upon the bank; and hence we may infer with logical propriety, that the energy (as evinced in the effects) has been increasing at that point for years; and being, now, at its highest, must from this, forward, steadily diminish, until it shall have, ultimately, passed altogether away. Below Wilkerson's, the testimony of Messrs. Offutt, Wallace, and Harris, shows an increasing energy in the increasing effect; and, therefore, as far as the irregularity of the outline, and the resistance of the soils will admit of a strictly exact result in such a case, *demonstrating* the present centre of effect to be *below* the 'Cut-off,' leads irresistibly to the inference that the time to speak of the suggested 'Cut-off' as within the limits of probability, has passed away. What the maximum impact failed to accomplish cannot be expected from a minor impact; nor is there any irregularity in the general outline of the bank to direct a special current against the debouch of the suggested 'Cut-off;' that outline, being in general a regular curve, may be held to receive, in the consequent uniformity of its resistance, an effect equally distributed. The rate of caving at the supposed 'Cut-off,' proves that the bank at that point is not inferior in cohesive strength, to that at any other point included in the information obtained in the case. Besides, the result suggested must now, if it come at all, come from *one* side; for the Eastern debouch of the 'Cut-off' has a making bank. If, then, the 'Cut-off' is to result from its present rate of caving, it will not, unless under some new and special combination of causes, occur for upwards of a century and a half. This supposes the centre of effect *constant* in its point of application; but with the centre *traveling* steadily to the Southward, the accomplishment of such a result must be deferred indefinitely. To sum up these remarks on the suggested 'Cut-off:' if the facts of the case do not positively establish that the 'Cut-off' will not be made, they go far to *prove* that such a supposition is entirely improbable."

The Report still further sketches out the method of reasoning, from the observed facts of "Making," and "Caving," in the following consideration of the question of increased shoaling at Gaines' Landing:—

"To meet your fourth question broadly, I must consider what other changes, as the supposition of the Cut-off must clearly be rejected, is most likely to take place in the River between Ferguson's Point and Gaines' Landing. The alignment of the River above the Railroad Terminus shows, as detailed above, a change of course, in a distance of three and a half-miles, of ninety degrees: in other words, the Mississippi River, curving from a point about three-quarters of a mile above the Railroad Terminus until it fronts the house of Mr. W. C. Campbell—a distance of three and a half miles—turns fully one-quarter round. To divert the whole volume of the Mississippi River so far from its direct course, implies the expenditure by the River of an immense energy on the resistance causing this divergence; and hence may we understand, in a general way, the amount of the force employed in operating on the bank between Mr. Campbell's plantation and the site of the Railroad Terminus. The caving consequent on the force so exerted against the bank between those points, stands at present in its progress to the southward, as follows : It begins at a point 500 yards below Mr. Campbell's ; and extending down the River-bank to the head of Island No. 80, a point 1200 yards below the Railroad depot, the centre of impact (the point of greatest effect) being at the present time situated, therefore, upwards of 2700 yards *higher up-stream* than the Railroad Terminus. The rate of effect at points along this bank I am unable to say ; but the maximum effect having yet to operate over a space of 2700 yards before it shall have reached the Terminus, has yet, in obedience to an infallible law of the River, to come sweeping down with all its powers of change and destruction on what remains of Ferguson's Point. In the march down-stream of the axis of current lies the cause of any such change of channel as may be looked for between the Railroad Terminus and Gaines' Landing. When the current of the River first swept the Northern bank of Ferguson's Point, the Southern bank of that point lay at the head of a line of slack-water. Island 80 resulted from this ; for the matter that passes off in suspension under the impulse of a current of 4 or 5 miles an hour, will be precipitated in currents of one or two miles an hour. Now, however, Ferguson's Point has been to a considerable extent carried away, within the last six years, to an extent, according to Col. B. Gaines and Mr. Reinhart, of eighty yards ; and as a consequence, the Island formed under the shelter of that Point begins now to receive the shock of the river current." Observations and applications of the above description being employed as a guide in the case of the location of Levees, the determination of those locations may be made with a proper regard to the most important considerations affecting their permanence. All points of the bank being thus examined under the light of the circumstances affecting their permanence, the limits of permanence inferred therefrom, must be noted at intervals on the plan; and the alignment of the Levee being made to conform to the considerations proper to itself, the location must be laid down on the plan within the restrictions of these limits of permanence. The laws governing the alignment of water-embankments, like those governing the alignment of Railroad tracks, point in the first instance to straight lines. The course of motion, whether of solids or fluids, is naturally rectilineal. As has been observed in the latter of the two foregoing extracts from the report of Mr. Hewson, the diversion of motion from its original line to any other line, involves the expenditure of more or less mechanical effect. In diverting a surface layer of the Mississippi flood-water—that mass moving at the rate of some 6 miles an hour—from one course to another, it can be readily understood that the expenditure of mechanical effect is very great. In order, then, to discharge this unavoidable force with the least possible danger to the Levee, it should (so that it be distributed *equally* over a large space) be discharged invariably over a *curve*. These few simple principles point out clearly the rules governing Levee alignment—straight lines where such are practicable, and regular curves where they are not. Laying down this curvilinear rectilinear alignment in a manner as far as possible to accord with the general lines of the river-currents, the Levee will be exposed at all its points to the *least possible* shocks and washes. The *limits of permanence* laid down on the plan according to the considerations premised above, the lines of current controling the general direction of the alignment, that alignment—making all its changes of direction over regular curves—may be laid down finally on the plan with the fullest faith in it as the location of greatest safety and greatest economy. Often, however, it will occur in reasoning on the considerations guiding in laying down the Levee-route on the plan, that two or more routes may *appear* to possess equal merits. Laying down all these routes on the map, each must be made a subject of instrumentation and estimation; and always taking into consideration that the closer the alignment adheres to the limits of permanence the greater the amount of good to the public, the relative cost of the respective routes determining, finally, as to the one for adoption. So much then for the general considerations affecting location. Special considerations in reference to stretches of considerably heavy embankment, may apply—such, for instance, as ridges furnishing, *within the limits of permanence*, an economical location for the Leveeing of a neck of swamp. These must in all such cases be examined carefully —first by the reconnoisance of a professional eye, and next, if found necessary, by instrumentation and estimation. So much then for the considerations applicable to location under the cramped action of the Mississippi Levee-laws.

CHAPTER VII.

SURVEYS.

High Water Mark, it has been shown, cannot be obtained so readily as is supposed by the populace. On the contrary, the correct determination of the flood-line for fixing the height of a Levee, is a duty that involves, not only sound judgment and patient investigation; but also careful and extensive instrumentation. The location of a Levee, it has also been pointed out, is something more than a matter of off-hand expediency. This duty of the Levee-system is at present—like the determination of the flood-line—assumed popularly to be fully within the knowledge and capacity of *every man* living on the banks of the river. The considerations affecting the discharge of such a task have, however, been shown already to be too intricate, too extensive, too delicate, to be grouped and combined into correct results by even men of fair standing amongst the members of the profession as Field Engineers. Location, with the commonest pretensions to care and science, requires, as has been indicated in the remarks on that head, as a first necessity, a full careful survey, an exact and *special* map. The first duty then of an improved system of Leveeing should be the preparation of maps and profiles—the surveys for those maps and profiles to be extended from end to end of those sections of country referred to hereafter as Drainage Districts. These surveys can be directed only by a mind quick in observation and ingenious in inference—this quickness and ingenuity guided by a familiarity with fluid-motion and river phenomena. They should show by actual offset-chaining the line of bank; and by careful sketching, all "makes," "bars," and currents. These instrumentations should bring out all the facts of cavings, so as to furnish to the mapper the penetration, progress, and stage of each cave. All facts of possible or probable influence on the objects of the survey—such for instance as the facts of Moon Lake in Mississippi, of Old Town Lake in Arkansas, of Bayou Atchafalya in Louisiana, their position, form, level, flow, &c. &c—ought to be carefully ascertained and connected with the great body of the facts of the District survey. Every foot of survey, whether of experimental lines along ridges, across swamps, or any where else, within a Drainage District, should be laid down regularly when completed and connected with the general survey on the plans and profiles of that District. These plans should consist of two sets; one set on a scale as large as practicable for a map of convenient size, showing the ground along the whole front of its whole District. Divided into squares by light lines across its face, this map should be made an index map by numbers set on each square so shown, to the several sheets of the second set of maps—a set made to a sufficiently large scale to embody all the minutiæ necessary for practical purposes. These enlarged working-plans, amongst the other particulars referred to as guides in location, should show the site and title of all survey-stations, the site and number of all Bench Marks, the elevations of the Bench Marks recorded duly by transfer from the District profile. The first exact and minute survey of a Drainage District effected by a special staff, the constant staff required for the Engineering direction of the District-works, should spend all the spare time from construction-duties, in keeping up, by survey, connected records, on the working plans, of all changes of "bars," increases of "makes," shiftings of currents, penetrations and progressions of "caves." These facts ascertained and laid down on the plans, year after year, the continuity of the records on the whole river will, after a time, enable a Levee-administration to reduce to something like scientific exactness, every consideration affecting the perfect practical efficiency of their most important duties.

The maps described here have been deduced as necessities of location from the circumstances affecting it on *but one side* of the river. But it has already been shadowed out in the remark on that subject, that the location of a work on *either* side cannot be made with *complete* care without the exact comparison with the location on the side opposite. The practical difficulties referred to, under the head of High Water Mark, also suggest the comparison of levels on one side of the river, with levels on the other side. But the necessity arising from these considerations is indicated still more forcibly from another point of view. The remarks offered on location show the necessity of avoiding all causes of excessive pressures, or shocks upon river-embankments. The currents treated with disregard, and the lines of least resistance duly observed, in location of a Levee, the conditions of location in reference to shocks are fully met, so far as the considerations affecting them on *that particular side* of the river. But let it be assumed that the Levees up-stream have, on both sides of the river, a considerable breadth of fore-shore; while at the point of this *locally* judicious location, the Levee on both sides happen to have for their fore-shore, each but a narrow strip. The width from Levee to Levee, across the river, may thus happen, up-stream, to be large, while below—at the point of the *locally* good location—the width across the river from Levee to Levee may happen to be comparatively narrow. This sudden contraction of the flood-flow will throw an increased shock of current on the Levees at the point of that contraction; and thus does the location of a Levee on one side, without due regard to that of the Levee on the other side, involve some of those *avoidable* contingencies of breaching the embankments which judicious location undertakes to guard against. Proper location, then, notwithstanding conformity with all considerations of "cave," current, and alignment, *on one side* of the river, cannot be made without comparing the location based on all these, with the location on the other side. The narrowest width of the riverflow, in the natural state, is said, in the late able pamphlet of Col. Alcorn, to be opposite Randolph, in Tennessee. A bluff at one side and a high bank on the other side, it appears that at that place the floods of the Mississippi pass off, without any particular increase of current, or wear of the bank, *within a width of* 2,000 *yards*. A proper survey of the river might probably throw further and more correct light on this particular fact; but whether Randolph be, or be not the site, and, whether 2,000 yards be, or be not the width, of the narrowest natural channel of flood-water, some site and some width answering those conditions ought to be ascertained for fixing *the ruling width* of water-way between the lines of river-embankment. This ruling width determined in reference to the width, section, and current of *several* "narrows" in the flood-flow, the proper location of Levees on either side of the river, requiring that the flood-width be never lower than the standard, such a location *on one side* can be made only *pari-passu* with the corresponding location *on the other side*. An inter-littoral survey is seen thus to be a necessity of economic and permanent location. This survey connecting District surveys across the river, does not require absolutely to be one of detail. Intermediate Islands should certainly be embraced in it; but in consideration of the cost of such an extension of labor, it is, perhaps, better (for some time at least) to omit soundings. A skeleton Trigonometrical survey, then, connecting stations in local surveys on both shores, and on intervening islands, is all that is absolutely necessary in addition to the surveys already described for completing the enquiries and records necessary to a perfectly correct and economic system of Levee-administration. The triangulation necessary for this survey, should be carried out with a view to fixing each station under the endorsement of one or more checks; but due regard to be paid, in all cases, to the regularity of the shape of the triangles, and to the including in each station-book on the field, of each of the stations that may be possibly combined in any one triangle. The correction of bases, the adjustment, in estimation, of spherical excess, &c., are details that, in addition to all the care suggested for the field, are highly necessary in carrying a base line of some 2000 or 3000 yards, with all the corrections of even several intermediate checks, through a series of some eight or ten hundred triangles. The triangulation, however, "poled out," the angles taken, the base measured, and the calculations made, the District-surveys may be carried out in detail as described, connecting regularly with the stations of the triangulation. The diagram of the trigonometrical points having been laid down, the filling in of this diagram, *on each side*, with the details of each local survey, will not only guarantee an accuracy *otherwise unobtainable in that local survey*, but will also present a perfect connection of the facts on *both sides* of the river. This connecting survey will, in the first place, by doubling the *data*, lead to reliable inferences in all cases as to the height of High Water Mark—will, by embracing in exact detail the facts of all the "narrows," limiting the width of flood-flow, lead to correct deductions as to the "ruling" width proper in the case of opposite Levees; and by representing the *relative* position of Levee-alignment on each side of the stream, point to those modifications or changes of site that may be necessary for conformity with the conditions of efficiency and permanence.

CHAPTER VIII.

ADMINISTRATION. *

The subjects of flood-line, location and survey involve necessities at evident conflict with the present system of Levee-legislation. In Arkansas and in Louisiana the administration of the drainage-interests are in the charge of the State; in Mississippi, in Tennessee, and in Missouri, the charge of those interests is parcelled out among the River-counties. In all these

---

* The opinions put forth here are found to be strikingly coincident with those of the Chief Commissioner of the Levees of Mississippi. His Report for 1856, to the Legislature of that State, has just been brought under the notice of the writer of this, and presents an opportunity for the endorsement of the views given under the above head, as in the following extract from that Report by so well-informed and judicious an observer:

"The practical results of the law placing the direction of the Levee within the respective limits of each County on the river, in the hands of a Board constituted on the principle of local representation, have been, so far as those results have fallen under my observation, decidedly unfavorable to the law. The act substituting a single Commissioner for these Boards of Commissioners in Tunica and Coahoma, has worked, in my opinion, much more advantageously to the interests of the Levee.

"This individual management is, in truth, in more close conformity with the physical principle that should direct legislation in this great practical work. No mere municipal line can divide an interest which is declared one and indivisible by the eternal law that rolls out the floods of the Mississippi in an unbroken whole. In not only principle, but also in practice, do I find reason to recommend this system of individual control in the design and construction of our Levee. It went into operation in the County of Coahoma two years ago, receiving from the previous recases the legislation is injudicious in its working—in Arkansas and Louisiana less so, however, than in Tennessee, Missouri, and Mississippi. The latter States presenting the extreme form of objection to non-conformity of Levee-law with Levee-requirement, the following remarks on points of this non-conformity are confined to the legislation of those States. The experience that has lead to the preparation of these remarks, has been acquired in Arkansas, and in Mississippi; and as the latter is one of the

gime, the legacy of a wasted resource, an exhausted treasury, an unsettled indebtedness, an imperfect record, an insufficient and incomplete Levee, and last, but worst of all, an almost total wreck of public confidence in any municipal administration. But what now, in two short years, is the condition of those affairs? Though I discuss a principle only in this case, it is not for me to answer, nor is my answer necessary when the answer has been already given in general terms by the County. This principle of individual management in carrying out our Levee has, in a direct issue with the principle of divided management, been endorsed emphatically by the intelligent people of Coahoma. Aware that the unity of the Levee could not be broken by municipal divisions, I had the honor to bring forward, two years ago, the existing law, giving a general jurisdiction over the Levee to the 'Superior Board of Levee Commissioners.' The working results of this law have fallen short of the physical principle which was sought to be reached by it. The interest of the river Counties is in truth such a perfect unit in reference to the Levee, more or less difficulty will always be found in carrying out so absolute a unity, under even the strongest organization of independent jurisdiction. A breach in the Levee at the upper end of Issaquena County, would, in the event of overflow through that breach, cause the destruction of property in the County of Washington by back-water. An overflow through the Levee at the lower line of Bolivar County, while it may do very little damage in Bolivar, may spread out one great sheet over the length and breadth of Washington County. In Tunica, an active caving of the river bank has already advanced within some fifty yards of the Levee, and still advancing, the next flood in the Mississippi will, in all probability, break in an immense volume into Eagle Lake. Now, to the greater portion of the people of Tunica, this result is a matter of comparative indifference—whereas the outfall from Eagle Lake, being Southwardly and Westwardly, such a result will spread devastation far and wide in Coahoma; so the construction of the Levee in the Southern border of De Soto, is a matter to the people of that County of comparative indifference—the majority interest is already provided for—the Levee is left open, and the country South of them becomes the sufferer. The local administration is the supreme power over that portion of the river

States whose Levee-legislation illustrates its conflict with Levee-expediences most forcibly, it is therefore selected here to illustrate that fact by examples mainly special to itself. The conclusions, however, though drawn to some extent for special instances, are general in their application—to those States where Levee-administration is distended to the extent of a whole State, and also to those where it is narrowed down to the limits of a County.

Drainage-legislation is based on error in limiting the admin-

within the limits of De Soto. Tunica has not the protection of a representation of the common interest which is bound up in the Levee an indivisible unit. Again, the Leveeing of those heads of Sunflower which traverse Lewis' Swamp, in the County of Coahoma, is a work of secondary concern to the great majority of the people of that County, but though situated within a jurisdiction regarding it with comparative indifference, this part of the Levee is of much deeper importance to the upper portion of Bolivar, than any like distance of low bank on her own front. While Coahoma required outlays at other points of much more urgency to her safety, her resources have naturally been employed at those points to the consequent injury of an immense amount of property in a neighboring jurisdiction.

"Indeed, such has been the interest felt in Bolivar in regard to this Levee, that influential citizens of that County, had offered, in addition to the only resources which Coahoma could agree to apply to that purpose, to pay a large bonus to any contractors, who would bind themselves in a contract with Coahoma, to Levee Lewis' Swamp. But, if a flood shall have risen before this swamp is Leveed, under the present state of affairs, how bitterly will the people of Bolivar regret, that while the local interests in the Levee have been provided for by an authority and an administration, there is no head, no strong individuality of general management, to represent the strong individuality of general interest. Wise legislation on practical improvements must always conform to physical laws. A general controlling authority is necessary also in this point of view, to represent the great and wide considerations involved in the intelligent design, and the straightforward independence, required in the faithful execution of that design, from end to end of that great physical unit, the Levee of the Mississippi and Yazoo bottom. From the commencement of the system, I have sought to convince the Levee interest of the necessity of this individuality; thus far my efforts have been unavailing. The plan of operation is one that I have never approved. I have been driven to its support for the reason, that no other plan could be suggested which could command the united support of the interests involved."

istration under it by arbitrary lines. In Shelby County, Tennessee, the proper administration of the Levees is not placed under the guarantee of any considerable interest. Some eight or ten thousand acres of swamp subject to the overflows of Nonconnah Creek and Horn Lake must always constitute an insufficient interest for the enforcement of an independent administrator of the Drainage-works of that area in the construction, protection, and maintenance of the Levees under his control—Levees extending to a length of some 15 miles. Indeed a question presents itself this moment as to whether, within the section referred to, there exists a single plantation, there resides permanently even a solitary squatter. The fact is, the Leveeing of the tract in question cannot, in all likelihood, be said under existing legislation to be the business of any one; but even if it be the business of any one, the area to be enclosed does not present, in all probability, the ways and means for raising—does not in short present a sufficient inducement to justify—the considerable expenditure required for its embankment—a sum that cannot be less than some $30,000. And yet, if this part of the bottom be left unenclosed, the whole Levee from the Tennessee line to the Yazoo, can be saved from utter uselessness for the drainage of the Valley but by a *special work* pressing on the limited resources of the Levee-interests in De Soto! If a physical facility have not brought this special embankment within the limits of the ability of De Soto, the inundations from Horn Lake will ignore the hamperings of Tennessee and Mississippi legislation by forcing *combined action of all the counties* between the Nonconnah and the Yazoo, in the construction, protection, and maintenance, of either a general Levee from the Tennessee line to the Nonconnah hills, or of a special embankment in De Soto County from the existing River-Levee to the Coldwater high lands. But instances of the bad adaptation of the present law are numerous. In Tunica County, Mississippi, the Commissioner is charged with several

Keys of the drainage of Coahoma, Sunflower, Tallahatchee—that at Buck Island Bayou, that at Couple-Timber Bayou &c. These, however, it may be said are Keys also to the Drainage of Tunica itself; and, therefore, are their safe-keeping placed in the hands of the local Commissioners under *some* guarantee. Tunica, however, is charged with another Key to the Drainage of Coahoma, Sunflower, and Tallahatchee—the Levee immediately covering the plantations on the North shore of Moon Lake. This latter Levee protects little or none of the settlements in Tunica; whereas the flood-water rushing through a crevasse therein sweeping southwardly across Moon Lake and the Yazoo, will inundate the fields and homesteads of Tallahatchee, Sunflower, and Coahoma. Want of interest in its construction, want of funds to pay for that construction, demands on their treasury and attention at points of concern to *themselves*, may lead the people of Tunica at any moment to regard this Key to the Drainage of Coahoma and its adjoining counties, with a very natural, and indeed quite excusable neglect. The most vital interests, then, of Coahoma, Sunflower, and Tallahatchee, are placed, by the system of County-jurisdiction in Leveeing, *beyond the control of these counties*—placed in the hands of parties who can afford without loss, to regard the protection of those interests with indifference. A tax to be collected from *themselves* for the construction or repair of the Levee covering Moon Lake—protecting *Coahoma*—would, naturally enough, be not carried probably without some effort amongst the people of Tunica. But the lower Counties show the working of the system of local-jurisdiction in still more objectionable lights. In Coahoma County there may be said to be no settlement south of Lewis' Swamp. The Coahoma people, as a body, care very little therefore, about the Leveeing of Lewis' Swamp; whereas, the floods breaking through that swamp, may at any time after the failure of its Levee, inundate at even ordinary floods, the lands and homes of Bolivar and Sunflower, unless

Levee-jurisdiction be regulated by some limits more *practical* in their operation than those of arbitrary lines. The hampered workings of Levee-legislation are thus seen by a few illustrations to be unjust and unsafe for the whole Valley of the Yazoo— for De Soto, for Tunica, for Coahoma, for Bolivar, for Sunflower, for Tallahatchee, and (the contingencies of local indifference, local urgencies, and local taxation, accumulating unfavorably as the testing of this legislation is carried down-stream) the injustice and unsafety is still greater in Washington than in Bolivar; and as compared with Washington is still greater in Issaquena.

But what is the remedy for the evils of the present system of Levee-government? An extension of Levee-jurisdiction according to certain physical proprieties. Working necessities point clearly to the removal of the existing limits on the administration, in the State of Mississippi, of river embankments. The location considerations referred to above, operate in full force, whether or not the ground lie one-half in Washington, the other half in Issaquena. In locations so circumstanced the surveys to be made must be *common to both counties.* The flood-level too, is a subject of inquiry that, as shown above, cannot be cut short by a mere legislative fiction; and, here, again, is another point in which the practical duties of Levee-administration ignore the system of imaginary limits to Levee-jurisdiction. Other considerations point still more forcibly to the necessity of seeking some new boundaries for the limitations proper to that jurisdiction. Bolivar's voice and Sunflower's voice in the appointment of the Commissioner directing the Levees of Coahoma, will guarantee the construction, protection, and maintenance, of a Levee across Lewis' Swamp, quite as soon and quite as surely, as that across the Yazoo Pass, or at any other point in Coahoma. So also as between Coahoma and Tunica: * give Coahoma, Tallahatchee, Sun-

---

* The joint interest of conterminous counties in the proper administration of their respective Levees and in the making and skill of their respective surveys, is

flower, votes in the election of the administrator of Levees in Tunica; and Coahoma, Tallahatchee, Sunflower, will assuredly be thereafter saved from the dangers, the, perhaps, ruinous indifference that, under the present law, may at any moment inundate their hearths and fields by overflows discharged upon them in desolating volumes through Moon Lake. But what distribution of jurisdiction will conform best to the practical and social considerations entering into Levee-administration? From Cape Girardeau in Missouri, where the highlands abut upon the river, to the mouth of the St. Francis in Arkansas, where the back-drainage of the intervening country must be discharged, defines a Levee-district, which, bound together by a community of interest, is for all the purposes of proper Levee-administration, an absolute unity. From Sterling in Arkansas, at the mouth of the St. Francis, where the Levee rests on the slopes of Crowley's Ridge, to the mouth of White River, where the back-drainage of the intervening country discharges, is also, so far as the Mississippi Levees are concerned, a unit in Levee-interest; and, therefore, should be a unit

pointed out by inference in the following remarks of Judge Hardeman, as Levee-Commissioner to Tunica County: "The object of making this survey was, with the then indication of the active caving of the bank of the River at the Southern part of Trotter's Field, to ascertain whether for the protection of the back-country, we would be compelled to Levee around Eagle Lake. *** The indication of cave, &c., which may save this county and *Coahoma* a considerable amount of money, &c. The question, however, as to whether we go around Eagle Lake ought to be determined by concert of action between the Levee Commissioners of Tunica and Coahoma, as they may decide as to the best interests of the two counties; for, as before remarked, there is a common interest of the two counties in erecting a Levee across the Pass, &c." Report for 1856, page 5. This extract specifies an instance in which the location of a Levee in one Levee-jurisdiction is held to involve a loss or a gain of a "considerable amount" of money to the tax-payers of another Levee-jurisdiction; and specifies also an instance in which the construction of a Levee within one jurisdiction is considered to be a question of drainage within another jurisdiction. And the parties thus concerned, in the one case in their pockets and in the other case in their property, to be deprived of all influence in making that location, or in expediting that construction.

in Levee-administration. From Pine Bluff, the nearest escarpment to the Mississippi of the Arkansas Uplands, to the mouth of Red River in Louisiana—the outlet of the rain-shed of the intervening bottom lands—the community of interest in the inclosing Levee is so indissoluble that the proper administration of that Levee over-riding all imaginary boundaries—whether of County or of State—must, in furtherance of sound policy and capable management, be centered of necessity in one and the same intelligence. On the Eastern bank it has already been indicated sufficiently plainly, that the Levee-interests from the base of the hills below Memphis in Tennessee to the mouth of the Yazoo River—the debouch-channel of the back-drainage of the included area—is so thoroughly identical in its drainage affairs—socially and practically—that the administration of those affairs within that area has been described by Col. Alcorn, most correctly, as "one and indivisible." The natural—the social and the working—definitions of the remaining jurisdictions in Louisiana and in Mississippi, may, with the views presented above, be determined by those acquainted with the physics of those sections; and so also of the jurisdictions in Tennessee, Missouri and Illinois. The limits assigned the districts defined here are, it ought to be observed, those on their Mississippi front, the limits on their inland side, in each case being located as hereafter indicated on such lines as may be necessary for the equal distribution of Levee taxation. Sufficient, however, has been said here to show that physical considerations applied socially and practically, while *ignoring the existing limits* to Levee-administration, describe plainly certain limits demanded for its efficiency. *

---

* Some four years ago the grounds taken here were taken by Col. Alcorn, see note to page 121, in urging the consolidation of Levee-government in his Report as chairman of the Superior Board of Levee-Commissioners, to the then Legislature of Mississippi. The essential unity of Levee-management was suggested subsequently

Unity of interest can be served truly by only administrative unity. Each *Drainage District* then (as the united areas referred to above may be termed,) ought to be placed under a single administrator. One Commissioner should be charged with the direction of drainage embankments from Cape Girardeau to the mouth of the St. Francis; one from the mouth of the St. Francis to that of Arkansas River; another from Pine Bluff to the mouth of Red River. On the opposite side a single Commissioner should be charged with the control of the Levee-interests from the Nonconnah to the debouch of the Yazoo. But while the social and the practical considerations in the case conspire

by Judge Hardeman, Levee-Commissioner of Tunica in his report for 1856, to his fellow commissioners, Messrs. E. B. Bridges and J. A. Cole, in the following judicious remarks: "The propriety of this repeal may be a question of doubtful policy, as it must be apparent to you that *a common interest in the Levees fronting all the counties on the Mississippi River from Horn Lake to the mouth of Yazoo River* ought to be appreciated by all land-holders within that Delta formed by Coldwater, Tallahatchee, and Yazoo Rivers, to its entrance with the Mississippi, a distance of 350 miles, embracing a part of the County of De Soto, all of Tunica, Coahoma, Bolivar, Washington, Issaquena, Sunflower, and part of the counties of Warren, Yazoo, Tallahatchee, and Panola. * * * Tunica County tax-payers on lands bordering on Coldwater, are as much interested in the Leveeing around Horn Lake; De Soto County, as they are in the Levee of their own County fronting the Mississippi River; also the land-owners bordering on the Mississippi River and Coldwater are equally interested in Leveeing the Yazoo Pass in Coahoma, as the water in making its way through the Pass backs up through Moon Lake, &c., to the town of Austin and its vicinity. The question may be well asked: *can this common interest in the Levees on the river be carried on without concert of action, &c.*" Messrs. Hardeman, Bridges, and Cole, are gentlemen of intelligence, of practical acquaintance with the working of the Levee system; and, as such, their testimony to the fact of Levee-unity, so far as it goes, is highly valuable. The remarks in the above extract point to the restoration of the Superior Board of Commissioners as fulfilling all the suggestions of Levee-concert; but, loose and scattered in its parts, the action of that Board has already been found to be utterly inefficient. Some other form of government, therefore, must be instituted to meet the *universally accepted fact* of Levee-unity; and the form of a general Board having been tried and found wanting, the necessities of the moment point to the only practical—indeed the only untried—form remaining, that of **administrative individuality.**

to define the working limits of District Levee-jurisdictions, the practical considerations point to a conclusion still in advance of existing systems. The working expediences involved in capable administration of Levee-drainage are not confined to one side of the Mississippi. The comparison of High Water data obtainable on the East bank with that obtainable on the West bank, has been referred to as an expediency in determining the question of High Water Mark. This comparison, then, indicates the extension of District-administration to an administration of wider scope and more general duties. The necessity of full knowledge of the location-facts on the opposite side, and of certain accord between the locations on both sides, is another instance under a system of District-management of a commingling that leads plainly to a further widening of administrative community. A fusion of District units is an expediency on these grounds; and therefore, on the further ground that, only by such a fusion can the Levee-interest of the great Valley of the Mississippi receive the first great contribution, the prime essential, of a broad, capable administration—a full, correct and *connected* set of working maps. The scientific and practical conditions of the drainage of the Valley by Levees require, therefore, that the administration of each Drainage District to the full extent of its *natural* limits, be placed in the hands of an *individual Commissioner;* and further require that the administration of all joint-duties of the Drainage Districts on both sides of the river, be placed in the hands of *those individual Commissioners assembled in general Board or Council.* Legislation based on the system of administration sketched out here, is clearly the only one adapted to the direction of those important works under the lights of scientific principle, of practical forethought, of sound economy.

Great difficulties, however, obstruct the effective working of proper machinery for the managemert of Mississippi Levees. The popular *intelligence* holding the purse-strings of the system does not, in some cases, go to the extent of recognizing in Leveeing, any skill beyond that of its own crude observation. It sometimes commits the mistake of ignoring the existence of centres of special knowledge, whether in Medicine or in Engineering. Col. Alcorn has been constantly hampered in his Commission by this condition of public opinion. "Such," he says, in his last pamphlet, " is the disposition to economise, that complaints are made should the Commissioner employ an Engineer at a salary of fifteen hundred a year! The subject must be elevated above this, or decent men will cease to be connected with it." Laughable as such difficulties to proper administration as those indicated in this extract may appear, they present in practice serious embarrassment to intelligent and vigorous administration.* His intelligence, his personal

---

* One of the embarrassments to the Levee-reformer, remaining as a consequence of the former employment of non-professional men for Engineering Levees, presents itself in the want of faith amongst even intelligent Planters of the Valley, in the skill and independence of the professional Engineer. Identified widely and favorably, as has been the name of Mr. M. Butt Hewson, with the leading measures of public improvement in the South-West, for several years—known, as it is, honorably to the professional Engineer, and the Railroad public generally, from New York to New Orleans—the Chairman of the Board of Levee-Commissioners for the State of Mississippi, was obliged, in 1855, to go into the defence hinted at in the following remarks from his Report of that year:—"The Messrs. Hewson, both M. Butt, the older, and William, the younger, are Civil Engineers by profession—have been schooled to the science—have, by competent men, been heretofore employed directing some of the most important public works of the South. Their labors have passed the ordeal of severe criticism; their competence has not been disputed by those qualified to judge. I cannot be required to stop and argue questions with men who oppose *their* calculations—who urge in opposition thereto, the figures of men who have emerged suddenly from the walks of private life, for convenience sake, to the dignity of Civil Engineers." Sound economy demands the employment of the very best men for the popular and for the professional duties of the Levee; and these once employed, administrative vigor demands that they be treated with the fullest confidence. Disparagement of the parties entrusted with those important duties, will merely weaken their hands, diminish their efficiency; and ought to be, therefore, frowned down by the intelligent and judicious, unless, when based on

pride, his honest conviction, and his whole property at stake, on the success of the Levees, an advanced man like the Commissioner for Coahoma ought not to be met, after the experience of the public in those works for fully seven years, with narrow and silly objections to his employment of an Engineer. Simple as the operation of shovelling and loosening earth undoubtedly is, the public in the Yazoo Valley, have not yet realized the fact, that even that simple operation is an important subject of practical science. Millions of dollars—national wealth, and national advancement—at stake on the shovelling, loosening, and hauling of earth, many of the tax-payers behind the Levee have yet to learn, or to value, the fact that even this item in Leveeing, taken from the blind guidance of the rude and wasteful suggestions of uninformed laborers, has been placed under the infallible guidance of economic inductions incorporated into a few practical laws. Leveeing, in fact, is in every particular an *art*. It requires more scientific skill, patient reflection, careful instrumentation, and, perhaps, even more practical knowledge of earth-works and foundations, than is required in the Engineering of nine out of ten of the Railroads of the country. Besides that, the Engineer entering on such duties, takes the position in his profession on this Continent, of the *pioneer of a new set of works*, the classifier of a new set of circumstances, the observer of a new set of phenomena; and consequently, to be professionally—that is to say, economically—successful as such pioneer, classifier, observer, must be guided from the outset by all the lights of the practice of water-works and of the science of fluids. Railroad Engineering is a beaten track. Uniform in almost all its details, that department of the profession involves, for the greater part, but a mere knowledge of routine rules. Routine practice, then, will constitute but a very poor qualification for a position, that like

unmistakable grounds that may be followed up to summary dismissal. No unfit man should be retained; no fit man should be—for the promotion of some petty interest—damaged in his efficiency.

Leveeing, must *make for itself its own rules of practice*—rules that cannot be made by ever so thorough a knowledge of mere routine, unaccompanied by a knowledge of the principles of practical and scientific Engineering. Thoughtful intelligence, then, appreciating the serious interests at stake in the efficiency of the Levees, instead of carping at the placing of the *professional* duties of the Levee-system in the hands of a regularly trained Engineer, in conjunction with the most able, enlightened, and honest man to be found for discharging the *popular* duties of the system, would rather have suggested its serious apprehension that his acceptance of a salary so small as $1500, place in a doubtful light the professional fitness of the Engineer charged with duties so delicate and responsible. The terrible lesson of the flood just subsiding, will, however, force the property and the purpose of the great Valley to action—action guided by all the lights of the broadest and most liberal intelligence. It is, therefore, hoped that, in order to sustain this expected action, some steps be set on foot for freeing Levee-administration from the popular drawbacks upon its efficiency, by either raising it as far as is practicable *above local restraints*, or, failing in that, by enabling the popular intelligence controlling the whole system, to keep pace with the growth of the importance of that system. *

* The views of administration presented above, have been endorsed in the late message of Governor McWillie, of Mississippi. In that able document, his Excellency holds the following language:—"This is a matter in which Mississippi is not alone interested, even on her own Levees. All the States, above and below her, along the river bank from Cairo down, are subject to the same inundation, and mutually act and react upon each other. The Levees of any one State are parts of a chain of Levees; and the direction, restraints, and flow, of the waters of the Mississippi through, or past any State, are portions of the forces which affect its regimen everywhere, but most strongly in the Counties below. * * * No elaborate plea is necessary to prove the importance of having a Levee-system for the whole Valley of the Mississippi, framed on sound principles of science, and in concert among the States interested." Change of administration—widening and concert of the several existing areas of administration—are felt on all hands to be necessities, and the best means of remedying this necessity practically is, undoubtedly, that of the District Drainage system extended to the organization of a general council.

CHAPTER IX.

EARTH WORK CALCULATIONS.

The Prismoidal Formula constitutes the only rule by which regularly sloped embankments can be measured correctly. This rule is as follows: *To the sectional area of each end add four times the area of the middle section; one-sixth of the resulting sum multiplied by the length of the prism, gives the solid content.* If lineal yards be the units employed in the calculation, the direct result of the rule will be cubic yards; but if lineal feet be the units employed in the calculation, then the direct result of therule being cubic feet, must, to express itself in cubic yards, be divided by 27. This formula supposes the embankment to be a regular prism; the actual height of crown and width of base midways between the two ends, being the arithmetical mean—one-half the sum—of the corresponding heights and widths, respectively, of the two ends. The end areas, then, must never exceed such limits as include between them observable inequalities of ground—the supposition of the rule being that the end areas have been taken at intervals sufficiently close to have broken the irregularities of the work into a series of uniformly sloping prisms. The "sectional area" referred to in the rule is the production of the arithmetical mean—half the sum—of the width of crown and width of base by the height. The 100 feet chain is that employed by Engineers in order by taking the end areas, whenever practicable at that distance asunder, to simplify the above prescribed rule for multiplying by the length. This multiplication in the case of stations 100 feet asunder is made by removing the decimal point two figures to the right; or when multiplying a whole number by the addition to it of two cyphers. Lineal feet being generally the unit of measurement, the prismoidal formula involves in general a division by 27 to reduce its result to cubic yards; and as it involves also a previous division by 6, it is somewhat of an abridgment when working it out in detail to divide the sum of the end areas and four times the middle area, after removing the decimal point two figures to the right, by 162—six times twenty-seven. In order to explain more clearly the working out of a measurement under the prismoidal formula, let it be required to calculate the number of cubic yards in a regular embankment 100 feet in length, ten feet high at one end, and 4 feet high at the other end, the crown having the uniform width of 5 feet, the base *of the slopes* being proportional to the height as six to one. The base for the end 10 feet high is (six times 10 for *the slopes* and 5 feet for *width of crown*) 65 feet; and one-half the sum—the arithmetical mean—of the width of base and width of crown being (65 and 5 divided by 2)˙35, the product of the arithmetical mean by the height (35 by 10) is 350, the area at the large end. By a like process the sectional area for the small end is 68. The arithmetical mean—one-half the sum of the heights at both ends—is (10 and 4 divided by 2) 7—the height of the middle section. The area corresponding to this height, by the calculation explained before, is 182; and this multiplied by 4 gives a product of 728—4 times the middle area. 350 (one end area) and 68 (the other end area) and 728 (4 times the middle area) show a total of 1146; and this multiplied by the length being 114600, one-sixth of the product divided by 27 (or 114600 divided by 162) shows a quotient of 707.40, the content of the embankment, in question, in cubic yards. To explain this more clearly it is better to repeat the same calculation in another form:

GREATER END.	LESS END.
10 (height)	4 (height)
6 rate of slope,	6 rate of slope,
60 base of slopes,	24 base of slopes,
5 base of crown,	5 base of crown,
65 width of base,	29 width of base,
5 width of crown,	5 width of crown,
2)70 sum of widths,	2)34 sum of widths,
35 arithmetical mean, or half the sum of the width of crown, and of base.	17 arithmetical mean, or half the sum of the width of crown, and of base.
35 arithmetical mean width,	17 arithmetical mean width,
10 height,	4 height,
350 sectional area.	68 sectional area.

MEAN AREA.

10 height of greater end,
4 height of less end,

2)14 sum of the two heights,

7 mean height.
6 rate of slope,

42 base of slopes,
5 base of crown,

47 width of base,
5 width of crown,

2)52 sum of widths,

26 arithmetical mean or half the sum of the width of crown and of base.
26 arithmetical mean width,
7 height,

182 sectional area,
4 multiplier,

728 four times the middle area.

SUMMATION.

350 area of greater end,
68 area of less end,
728 four times middle area

1146 aggregate of areas,
100 length,

6)114600 product of aggregate by length,

27)19100 content in cubic feet,

707.40 content in cubic yards.

This then is the full detailed method of earth-work estimation in accordance with the prismoidal formula. Before offering any further remarks on the subject, it is better to meet here the prevailing practice of estimation among the unskilled men charged with the "Engineering" of Levees, by some comparisons with the above—the correct—practice. The methods of calculation in common use on all the Levees of both Arkansas and Mississippi—with the exception of those in Coahoma, where Col. Alcorn has taken the trouble to acquire perfect facility in the correct practice himself—are those by *average heights* and by *average end areas*. These systems are wrong in principle; but, in the popular spirit by which these remarks have been guided, waving the error of principle, the most effective corrective in the case will be an illustration of that error in practice. The prismoidal formula has already been worked out in detailing the measurement of a Levee ten feet at one end and four feet high at the other end, the width of crown being uniformly 5 feet; and the aggregate rate of side slopes 6 to 1. The content of this embankment will now be detailed according to the two rules of measurement pursued generally on the Mississippi:

BY AVERAGE HEIGHTS.

```
 10  height at greater end,           182  area for average height,
  4  height of less end,              100  length,
 ──                                   ────
2)14  sum of heights,               27)18200  content in cubic feet,
 ──                                   ──────
  7  average height,                  674.00  content in cubic yards.
  6  rate of slope,
 ──                                   BY AVERAGE AREAS.
 42  base of slopes,
  5  base of crown,                   350  area of greater end,
 ──                                    68  area of less end,
 47  width of base,                   ────
  5  width of crown,                 2)418  sum of areas,
 ──                                    ──
2)52  sum of widths,                   209  average area,
 ──                                    100  length,
 26  half sum of widths.              ─────
  7  average height,                27)20900  content in cubic feet,
 ───                                  ──────
 182 area for average height,         774.00  content in cubic yards.
```

By this method of average heights, then, the solid content of the embankment in question, would be taken at 674 cubic yards; and by the method of average areas at 774 cubic yards; a difference that at fifteen cents a yard, showing a discrepancy of $15 for 100 feet of Levee, would sum up at the same rate to an immense sum when repeated for every 100 feet along the whole extent of even a County. But the fact of the case is; both of the quantities are wrong; and the one—that by average areas—being wrong in its *excess*, is an injustice to the tax-payer, while the other—that by average heights—being wrong in its *deficiency*, is an injustice to the contractor. The true quantity, as given in accordance with the prismoidal formula, has been shown in detail to be 707 cubic yards. The quack-systems then, and the correct system, compare in the case under consideration as follows:

   674 cubic yards—the content by average heights.
   707 cubic yards—the content in fact.
   774 cubic yards—the content by average area.

One of the common systems, then, of calculation by average heights is an injustice in the instance under consideration, at the rate of $261 per mile to the Levee contractor; the other and equally common system is an injustice in the same instance at the rate of $531 per mile to the Levee-tax-payer. This assumes the cost of the work at 15 cents per cubic yard. The error of these modes of calculation are sometimes less than in the case presented above; but they are also, sometimes, even still greater: with end-areas and end-heights nearly equal, they are very trifling; but with end-areas and end-heights differing largely, those errors become very serious. For the width of crown and rate of base adopted in the above example, the excess of result, according to the system of *average areas* for example, increases over the true content of the prism according to the following gradations: for lengths of 100 feet, where the inequality of the heights of the two ends is,

1 foot, the excess given by the method of average *areas*, is			1.6 cubic yards.
2 feet,	do	do	7.5 cubic yards.
3 feet,	do	do	17.0 cubic yards,
4 feet,	do	do	30.0 cubic yards.
5 feet,	do	do	46.0 cubic yards.
6 feet,	do	do	66.6 cubic yards.
7 feet,	do	do	90.7 cubic yards.
8 feet,	do	do	118.5 cubic yards.

Adopting still the five-feet crown and six-fold base, the insufficiency of the quantities resulting from the process of *average heights*, follows the following gradations: with a length of 100 feet where the inequality of the heights of the two ends is,

1 foot, the deficiency by the method of average *heights*, is			0.8 cubic yards.
2 feet,	do	do	3.7 cubic yards.
3 feet,	do	do	8.5 cubic yards.
4 feet,	do	do	15.0 cubic yards.
5 feet,	do	do	23.0 cubic yards.
6 feet,	do	do.	33.3 cubic yards.
7 feet,	do	do	45.4 cubic yards.
8 feet,	do	do	59.3 cubic yards.

It may be noted here that the calculations by *average heights* are always a wrong to the *contractor*, those by the average *area* being always a wrong to the *public*—the *deficiency* in the one case being one-half that of the *excess* in the other case. But, bad as are both of those methods when applied to even short lengths of embankment, a very common practice in the use of both, by extending the averages to *considerable lengths*, make the evil still greater. The following table shows the heights, crown, and base of a Levee, taken at regular intervals of 100 feet; extracted from a measurement-book of my own practice on Levees, it represents an actual state of facts. The last column shows the content of each 100 feet of the embankment, according to the prismoidal formula—the true content—in cubic yards.

Stakes.	Heights.	Widths.		Contents in cubic yards
		Crown.	Base.	
1210	3.36	5	20.16	
1211	2.59	5	15.54	124.9
1212	1.77	5	10.60	74.3
1213	1.60	5	9.60	47.5
1214	1.43	5	8.58	39.2
1215	1.80	5	10.80	43.7
1216	2.52	5	15.12	71.8
1217	1.74	5	10.44	71.4
1218	3.00	5	18.00	85.0
1219	3.72	5	22.32	156.1
1220	3.18	5	19.08	164.5
1221	3.91	5	23.46	173.6
1222	3.73	5	22.38	190.4
1223	3.63	5	21.78	186.3
1224	2.60	5	15.60	137.1
1225	1.96	5	11.76	79.9
1226	6.28	5	37.68	243.0
1227	9.40	5	56.40	706.2
1228	7.58	5	45.48	584·5
1229	12.38	5	74.28	1225.1
1230	11.26	5	67.56	1659.8
1231	8.69	5	52.14	1207.8
1232	10.23	5	61.38	1283.9
1233	8.66	5	51.96	1082.2
1234	7.66	5	45.96	814.4
1235	7.49	5	44.94	704.7
1236	8.50	5	51.00	786.1
1237	8.74	5	52.44	903.1
1238	7.47	5	44.82	805.6
1239	6.12	5	36.72	577.1
1240	3.04	5	18.24	285.3
1241	2.10	5	12.60	97.2
1242	1.75	5	10.50	60.0
1243	1.63	5	9.78	47.9
1244	1.76	5	10.56	47.9
1245	2.09	5	12.54	60.0
1246	2.22	5	13.32	71.4
1247	3.49	5	20.94	118.3
1248	3.47	5	20.82	167.9
1249	3.43	5	20.58	164.4
1250	3.42	5	20.52	160.7
	Total cubic yards,		· · ·	· 15,676.2

The true content of the above Levee—4,000 feet long—was 15676 cubic yards. Now, the average height of all the stations on this piece of work was, as may be seen by adding up the above column of heights, and dividing the sum by the number of heights so added, 4.67 feet. The average width corresponding to the average height, being 16.51 feet, the area—the product of this average height by its average width—is 77.10 square feet. To repeat this in another form:

 16.51 the average width corresponding to a height of 4.67 feet.
 4.67 the average height of the whole embankment.

 77.10 the average area by a general average height of the whole bank.
 4000 the length of the whole bank.

27)308400, content of whole bank in square feet.

 11422.22, content of whole bank in cubic yards.

The comparison in this instance then, stands thus:

    15672 cubic yards, the true content.
    11422 cubic yards, the content by a *general average* height.
    ———
      4250 cubic yards of error against contractor.

The contractor, in this instance, paid nominally, 15 cents per yard for his work, would, in fact, be paid according to this system of measurement, at the rate per yard of less than 11 cents. The contractor, however, is generally able to secure fair play for himself; but in the case of those methods of calculation that, pursued as they are by officers of the public, give *the contractor a large excess* above his *just rights*, there is no protection to save that public, when it contracted for but 15 cents a yard, from paying, in consequence of the unfitness of its own officer—its "Engineer"—so high, in fact, as even 20 cents a yard. So important is it to both the contractor on Levees, and to the public paying for their construction, that a system of measurement be laid down that, adapted to the popular understanding, may secure to both parties mutual, even-handed justice.

    The errors of the systems common in measuring Levees thus exposed, attention may be now recalled to the prismoidal rule. The illustration given of that rule will have suggested that its employment at intervals of 100 feet, and of less, along a line of Levee, makes correct estimation, a process most elaborate and tedious. Practice, however gives a surprising expertness in casting up quantities directly; and also in the use of regular forms of calculation, suggests from time to time several methods of abridgement. For a regular rate of slope, for instance, the Engineer about to estimate any considerable stretch of work, finds it much more correct and rapid to calculate, in the first place, a regular table for that slope; and applying that table to the special dimensions of his measurement, take off prismoid after prismoid, by inspection. For new Levees such a table is directly applicable. Tables 1, 2, and 3, have, accordingly, been added at the end of these remarks, for the use of the less expert—and indeed, also, of the more expert—to whom the elaboration necessary, otherwise, may be an obstacle to the general introduction in Levee-measurement of the prismoidal formula. These calculations are intended to cover all the forms of section prevailing in the Levee-practice of the Mississippi. In terms *representing cubic yards*, the tables show, for prisms of 100 feet long, the "end area," and "four times middle area" for all heights to tenths and half tenths of a foot, from a height of one foot to a height of 24.95 feet. Table No. 1 is estimated for a base-width of 6 feet horizontal to 1 foot vertical; and with a crown of 5 feet wide—the dimensions allowed by the Superior Board of Levee Commissioners for the State of Mississippi. Table No. 2 is estimated for a base, bearing the same constant proportion to the height; but differing from table No. 1 in having a crown of only 3 feet wide. No. 3 gives the quantities under the same heads, in the same terms, and for the same intervals, for a Levee having a crown of 3 feet across; but with a base having a width of *seven* times the height. The dimensions given in this table, are those generally used in the State of Arkansas, with the exception of the width of crown; the adopted crown-width being, as before remarked, erroneous in principle. In measuring the Levee it is, in fact, not practicable to arrive at a greater accuracy in the heights than a tenth of a foot. The tables are accordingly, in being extended to tenths, carried out to the fullest detail available in practical estimation. In order to show in juxta-position a calculation made in detail, and the same made under the abridging of the above table, let it be proposed to cast up the quantities in a Levee 100 feet long, 3.60 feet in height at the less end, and 7.70 feet in height at the greater end, the base being always *six* times the height, and the crown of the uniform width of 5 feet.

EMBANKING LANDS FROM RIVER-FLOODS.   143

PRISMOIDAL FORMULA WORKED OUT IN DETAIL.

GREATER END.
 7.70 height,
  6 multiple for base,

 46.20 width of base,
 5.00 width of crown,

2)51.20 sum of width,

 25.60 mean—or half sum of width
 7.70 height,

 179200
  1792

197.1200 sectional area.

LESS END.
 3.60 height,
  6 multiple for base,

 21.60 width of base,
 5.00 width of crown,

2)26.60 sum of width,

 13.30 mean—or half sum of width,
 3.60 height,

 79800
  399

 47.8800 sectional area.

MIDDLE AREA.
 7.70 height at greater end.
 3.60 height at less end,

2)11.30 sum of heights

 5.65 mean—or half sum of heights,
  6 multiple for base,

 33.90 width of base,

 5.00 width of crown.

2)38.90 sum of width.

 19.45 mean—or half sum of widths,
 5.65 mean height,

 9725
 11670
 9725

109.8925 middle area.
  4 multiple according to rule,

 439.57 four times middle area,
 197.12 area at greater end,
 47.88 area at lesser end.

6)684.57 sum of areas,

 114.095 one-sixth the sum of areas,
 100 length.

27)11409.5 (422.6 solid content in cubic yards,
  108

   60
   54

   69
   54

   155
   162

Such is the regular working out of this quantity in detail. Let it now be worked out by the tables. The crown being five feet and the base *six* times the height, the table to be employed in the case is No. 1. Turning then to No. 1, under the head of three feet and on the line corresponding to the decimal .60 in the margin, the tabular "end area" corresponding to 3.60 is found to be 29.6; under the head of 7 feet and on the line of the marginal decimal .70, the end area in the table is 121.7. Adding together 3.60 the height at one end, and 7.70 the height at the other end, the aggregate is 11.30; and this sum divided by 2 shows for the middle height, 5.65 feet. Turning again to the table, the tabular number under the head 5 feet, and on the line .65, is found in the column of "middle areas" to be 271.4. Adding 271.4 (four times the middle area) 29.6 (the area at the less end) and 121.7 (the area at the greater end) the total is 422.7. This explanation of the use of the tables thus given, the comparison with the above detail may be now commenced.

```
3.60—less height—tabular "end area" corresponding,      -   -    29.6
7.70—greater height—tabular "end area" corresponding,   -   -   121.7
                                                                ─────
2)11.30 sum of heights.
                                                                ─────
5.65—middle height—tabular "4 times middle area" corresponding, 271.4
                                                                ─────
    Solid content, by prismoidal formula, in cubic yards,  - -  422.7
```

The figures in each process show at a glance the facilities furnished by the table—the detail process requiring 208 figures, and tabular abridgement but 29 figures. The tables, then, may be held as reducing the time and labor of calculations by the prismoidal formula to one-seventh the time and labor necessary in carrying out that formula in detail. These tables are altogether new—the result of using the formula extensively when cutoff from an opportunity of reference to any other system of calculation by inspection. Original in every particular as they are, it is, perhaps, better to explain more fully than has been done in the foregoing comparison, the use and convenience of those tables. Passing to this explanation, it may be observed that the quantities employed in the table are fictitious, representing no *real* quantity, until the summation into a solid content, when they take the form of *cubic yards*. Extracting the heights at each station from the level-book, these are transferred, in the office, to the measurement-book in the following manner.—The column showing the distances between the several stations (see annexed form) are to be filled up with those distances, *leaving every second line* blank. The heights, respectively, corresponding to those distances are then transferred to the column of heights, each opposite its own distance, and consequently entered, like the distances, on *every second line*. The third column of the measurement-book is next filled in with the quantity constituting an arithmetical

Form of Measurement Book adapted to Tables Nos. I., II., and III.

Distance Station.	End Heights.	Mean Heights.	Tabular Number.	Contents in Cubic Yards.	Remarks.
0	5.00		54.1		
		5.10	224.2		
100	5.20		58.1	336.4	58.1 belonging to this prismoid is included therein, and belonging also to the following, is included in that, too.
		5.50	258.0		
200	5.80		71.8	337.4	
		5.55	262.4		
300	5.30		60.2	393.9	
		5.50	258.0		
400	5.70		69.0	337.2	
		6.45	348.0		
500	7.20		107.1	524.1	
		13.40	1412.8		
18 feet	19.60		741.7	294.0	Special calculation.

mean between each pair of heights entered in the second column—this mean being, of course, one-half the sum of its corresponding pair of end-heights. These mean heights are entered in the lines that had been left blank, when filling in the first and second columns; and thus occupy, in the measurement-book, a place between the two heights, from which each of them is deduced. The end-heights, and mean heights thus filled in and placed in proper position in the measurement-book, the calculator will next call to his assistance the earth-work tables. In doing this, it must be recollected that each set of heights in the tables, having corresponding to it two different sets of quantities—that in the column of "end-areas," and that in the column of "4 times the middle height"—the only certain mode of guarding against the use of one of these for the other, is to first take out the quantities under one head—those for the measurement-column of "end-heights," first; and *all these completed*, then take out the quantities under the other head—those for the measurement-column of "mean heights." The first "end-height," then, is 5.00. Turning to table No. 1, the eye rests on the *head*—in large characters—" 5 feet." Running down the margin, the decimal "00" is seen; and the "end-areas" under the heading "5 feet" on line "00" is found to be 54.0. Under the head, Tabular numbers of the measurement-book, this 54.0 is then entered on the line running across the book from the end-height 5.00. The heading "5 feet," being again used in the tables, the eye rests in the next place on the marginal decimal .20; and the "end-area" under the head "5 feet," corresponding to the decimal .20, is seen to be 58.1. The 58.1 is, then, entered in the column of "tabular numbers" of the measurement-book, on the line running across the book from the corresponding height of 5.20. So, also, with all the other "end-heights." These completed, the next duty is to take out the tabular numbers for the "mean heights." These, be it remembered, are found in the column "4 times middle area." The first mean height in the above form of measurement-book, is 5.10. Again, under the head "five feet," after running down the margin to the decimal .10, the eye rests on the tabular number corresponding to 5.10 in the column "4 times the middle area." This number is seen to be 224.2. Opposite to, and on the line running across the measurement-book from, "mean height" 5.10, this number, 224.2 is next entered in the column of tabular numbers. The mean heights are thus gone through, one after the other. The use of the tables ended, the next step is the summation. This must be done by grouping together each three quantities in the column of "tabular numbers"—always taking care that, *after* that quantity corresponding to the *first* distance the quantity corresponding in the column of "tabular numbers" to each distance, or to each "end height," shall be used in the additions *twice*—*once*, in addition to the two quantities *above* it, and *again*, in addition to the two quantities *below* it, in the measurement-book. The column "content in cubic yards" thus made out for each distinct prism, the addition of *all* completes the measurement. This supposes the stations, it will be observed, separated by uniform distances of 100 feet each. In irregular ground, however, the stations must be separated by irregular intervals—in bayous, for instance, it being often necessary to place them so close together as 4 or 5 feet. The measurement-book in such cases is filled, as shown between stations 4 and 5 in the above form, the prisms being made subjects of special calculations. These calculations may be made with a saving of time and trouble by adding the tabular numbers corresponding to its end-heights, and to its mean height as described for the 100 feet lengths; and multiplying the sum of these numbers by the length of the short prism in question; the removal of the decimal in the product, two figures to the *left*, will give the true content of that prism. Suppose, for illustration, a prismoid of 13 feet length, 7.20 feet at one end, and 19.60 feet at the other end. This has a mean height of (one-half the sum of its end-heights) 19.60 added to 7.20, and the sum divided by 2—13.40.

7.20—end height has a tabular "end area" of - - - 107.1
19.60—end height has a tabular end'area of - - - - 741.7
13.40—mean height has a tabular "4 times middle area" of - 1412.8

Total, - - 2261.6
Multiplied by length - - 13
Cubic yards in the 13 feet prism - - 294.00.8

This will save some trouble, as otherwise the calculation for the 13 feet must be made under the fórmula in *extenso*. The remaining quantities in the above form of measurement are obtained in the same manner as those of the foregoing explanations. The quantities of embankments having a crown of 3 feet across and a base of *six*-fold width, are to be calculated by table No. 2. A *seven*-fold width of base having a crown of 3 feet wide presents a section whose quantities must be calculated by table No. 3. The use of these two tables is precisely similar to that of table No. 1.

Tables 1, 2, and 3 are confined in their application to new or well preserved embankments. Old Levees, however, with worn crowns, hollowed sides, and spread bases, cannot be measured with any approach to truth by a rule based on a uniform width of crown and constant rate of base. Estimation under such circumstances can be made only by a series of careful cross-sectioning ; and in order, therefore, to meet this necessity of the present works in Levee management, a table of contents is added here on the basis of sectional areas. Table No. 4 aims at this object. Irregular works being the special subject for the use of this table, it may be necessary to observe that it is equally applicable to works of uniform sections. The rule for using table No. 4 is as follows : For lengths of 100 feet *add to the cubic yards corresponding in the table to each of the given end areas, 4 times the cubic yards corresponding to the mean of those two areas, and the sum will be the content of the bank in cubic yards.* Or another rule for using table No. 4 : *Add for lengths of* 100 *feet the two end areas to* 4 *times the mean of those two end areas, and the number corresponding in the table to the total of these is the content of the bank in cubic yards.* For shorter lengths than 100 feet, multiply the result in either of the above rules by the length, and changing the decimal point two figures to the left, the product is the content in cubic yards. The sectional areas in the tables are given, it may be observed, in square feet. Taking a few of the prismoids in the form of measurement-book already given, the sectional areas are as follows, as estimated by table No. 4:

Form of Measurement-Book for Table No. IV.

Stakes.	Height.	Widths.			Sectional Areas.	Mean Area.	Contents in Cubic Yards.
		Crown.	Base.	Average widths.			
0	5.00	5	30.00	17.50	87.50	90.81	
1	5.20	5	31.20	18.10	94.12		336.8
						104.77	
2	5.60	5	34.80	19.90	115.42		388.0
						106.47	
3	5.30	5	31.80	18.40	97.52		394.0
						104.62	
4	5.70	5	34.20	19.60	111.72		387.4

```
         First prismoid—less end area      87.50
                       greater end area    94.12
                    4 times middle area   363.24
                              Total       544.86
```

The solid content corresponding in table No. 4 to this aggregate " area" 544.86 is 336.4 cubic yards.

```
         Second prismoid—less end area—94.12
                       greater end area—115.42
                     4 times mean area—419.08
                              Total    628.62
```

Running the eye down the column "areas in square feet" of table No. 4, it rests on the large figures 620; and following down the column to the single figure 8, the number corresponding to 628 is seen to be 387.6. By the aid of the auxiliary table subjoined to table 4 the proportional value in cubic yards corresponding to the decimal of the areas, .62 is to be added to the solid content of the whole numbers. The cubic yards corresponding to the aggregate area of 628.62 shows, therefore, a solid content of 388.0 cubic yards. This explanation is sufficient to make the use of table No. 4 perfectly clear. For the particular measurements contemplated by this table it will be found a most valuable assistant to the calculator who aims at close and careful calculation.

All the tables in this book are new to the profession; tables 1, 2, and 3, being modifications from the practice of Mr. M. Butt Hewson; table No. 4 being, however, *purely original*. Tables Nos. 1, 2 and 3, respectively, condense in a small sheet containing 60 lines and 16 columns, a number of results in cubic yards that cannot, in the ordinary diagonal tables of earth-works, be given in a smaller space than that occupied by 480 lines and 480 columns. Earth-work tables in general, limit their facilities to heights of *full* feet; and therefore, tables 1, 2, and 3, annexed to this, while much more condensed in form and much more facile of reference, make a great advance in going into detail so minute as that involved in heights of feet and tenths of a foot. It is perhaps unnecessary to state that all these tables are equally applicable to cut and to bank, whether on Levee, Canal or Railroad.

In conclusion it may be added that these remarks, whether in theory or in practice, have been of necessity generalities. Engineering on Levees is in the crude state of those improvements, work for a man of some original observation, some original resources, some scientific and practical skill. Correct measurement will be brought by the remarks made above within the compass of men of intelligence under *ordinary circumstances*; but the special circumstances for even separating Levee-practice from general rules—spreading of base, sinkings of foundations, bulgings of sides, inundating of work-pits, &c.—make it necessary for proper estimation of those works that they be always placed in the charge of some one thoroughly conversant with the elementary principles of measurement. Location alone involves so many delicate and intricate considerations—these again involving so many serious if not fatal contingencies—as to require, superior to all rules of practice the eye and mind of a professional Engineer. The detail surveys suggested above are undertakings, too, that in the hands of even a *decent* pretender to professional ability will result in a simple waste of money. The Trigonometrical survey for connecting the Levees on both sides of the River is a duty from which (let unfitness be ever so ready to undertake it) even the regular Engineer, who has never directed his mind or his practice to such a system of survey, will be found in honor and self-consciousness to decline. Finally: if anything that has been said here shall further the interests of Levees, shall bring those works more thoroughly within the rules of art, shall strengthen the hands of the administrator entrusted with their charge, or shall correct errors of opinion on the part of planters and others hampering the intelligence of his aims, the writer shall have felt rewarded with the satisfaction of having left the impress of his experience in the great Valley after him as a *souvenir* for the benefit, in a greater or less degree, of the very highest interests of a generous people, amongst whom he has spent many a happy day of work and pleasure.

THE END.

www.ingramcontent.com/pod-product-compliance
Lightning Source LLC
Chambersburg PA
CBHW030345170426
43202CB00010B/1256